STRESS IN ASME PRESSURE VESSELS, BOILERS, AND NUCLEAR COMPONENTS

STRESS IN ASME PRESSURE VESSELS, BOILERS, AND NUCLEAR COMPONENTS

Maan H. Jawad, Ph.D., P.E.
Global Engineering & Technology, Inc.
Camas, WA, USA

WILEY

This edition first published 2018

Copyright © 2018, The American Society of Mechanical Engineers (ASME), 2 Park Avenue, New York, NY, 10016, USA (www.asme.org).

Published by John Wiley & Sons, Inc., Hoboken, New Jersey.

The right of Maan H. Jawad to be identified as the author of this work has been asserted in accordance with law.

Registered Office(s)

John Wiley & Sons, Inc., 111 River Street, Hoboken, NJ 07030, USA

John Wiley & Sons Ltd, The Atrium, Southern Gate, Chichester, West Sussex, PO19 8SQ, UK

Editorial Office

The Atrium, Southern Gate, Chichester, West Sussex, PO19 8SQ, UK

For details of our global editorial offices, customer services, and more information about Wiley products visit us at www.wiley.com.

Wiley also publishes its books in a variety of electronic formats and by print-on-demand. Some content that appears in standard print versions of this book may not be available in other formats.

Library of Congress Cataloging-in-Publication Data

Names: Jawad, Maan H., author.
Title: Stress in ASME pressure vessels, boilers and nuclear components / by
   Maan H. Jawad.
Description: First edition. | Hoboken, NJ : John Wiley & Sons, 2018. |
   Series: Wiley-ASME Press series | Includes bibliographical references and index. |
Identifiers: LCCN 2017018768 (print) | LCCN 2017036797 (ebook) | ISBN
   9781119259268 (pdf) | ISBN 9781119259275 (epub) | ISBN 9781119259282 (cloth)
Subjects: LCSH: Shells (Engineering) | Plates (Engineering) | Strains and stresses.
Classification: LCC TA660.S5 (ebook) | LCC TA660.S5 J39 2017 (print) | DDC
   624.1/776–dc23
LC record available at https://lccn.loc.gov/2017018768

Cover design by Wiley
Cover image: © I Verveer/Gettyimages

Set in 10/12pt Times by SPi Global, Pondicherry, India

10 9 8 7 6 5 4 3 2 1

Contents

Series Preface

The *Wiley-ASME Press Series in Mechanical Engineering* brings together two established leaders in mechanical engineering publishing to deliver high-quality, peer-reviewed books covering topics of current interest to engineers and researchers worldwide. The series publishes across the breadth of mechanical engineering, comprising research, design and development, and manufacturing. It includes monographs, references, and course texts. Prospective topics include emerging and advanced technologies in engineering design, computer-aided design, energy conversion and resources, heat transfer, manufacturing and processing, systems and devices, renewable energy, robotics, and biotechnology.

Acknowledgment

The author would like to thank Mr. Donald Lange of the CIC Group and Bernard Wicklein and Grace Fechter of the Nooter Corporation in St. Louis, Missouri, for their support. Special thanks are also given to Dr Chithranjan Nadarajah for providing the finite element analysis of the quadratic element in Chapter 11.

1

Membrane Theory of Shells of Revolution

1.1 Introduction

All thin cylindrical shells, spherical and ellipsoidal heads, and conical transition sections are generally analyzed and designed in accordance with the general membrane theory of shells of revolution. These components include those designed in accordance with the ASME pressure vessel code (Section VIII), boiler code (Section I), and nuclear code (Section III). Some adjustments are sometimes made to the calculated thicknesses when the ratio of radius to thickness is small or when other factors such as creep or plastic analysis enter into consideration. The effect of these factors is discussed in later chapters, whereas assumptions and derivation of the basic membrane equations needed to analyze shells of revolution due to various loading conditions are described here.

1.2 Basic Equations of Equilibrium

The membrane shell theory is used extensively in designing such structures as flat bottom tanks, pressure vessel components (Figure 1.1), and vessel heads. The membrane theory assumes that equilibrium in the shell is achieved by having the in-plane membrane forces resist all applied loads without any bending moments. The theory gives accurate results as long as the applied loads are distributed over a large area of the shell such as pressure and wind loads. The membrane forces by themselves cannot resist local concentrated loads. Bending moments are needed to resist such loads as discussed in Chapters 3 and 5. The basic assumptions made in deriving the membrane theory (Gibson 1965) are as follows:

1. The shell is homogeneous and isotropic.
2. The thickness of the shell is small compared with its radius of curvature.

Stress in ASME Pressure Vessels, Boilers, and Nuclear Components, First Edition. Maan H. Jawad.
© 2018, The American Society of Mechanical Engineers (ASME), 2 Park Avenue,
New York, NY, 10016, USA (www.asme.org). Published 2018 by John Wiley & Sons, Inc.

Figure 1.1   Pressure vessels. Source: Courtesy of the Nooter Corporation, St. Louis, MO.

3. The bending strains are negligible and only strains in the middle surface are considered.
4. The deflection of the shell due to applied loads is small.

In order to derive the governing equations for the membrane theory of shells, we need to define the shell geometry. The middle surface of a shell of constant thickness may be considered a surface of revolution. A surface of revolution is obtained by rotating a plane curve about an axis lying in the plane of the curve. This curve is called a meridian (Figure 1.2). Any point in the middle surface can be described first by specifying the meridian on which it is located and second by specifying a quantity, called a parallel circle, that varies along the meridian and is constant on a circle around the axis of the shell. The meridian is defined by the angle θ and the parallel circle by ϕ as shown in Figure 1.2.

Define r (Figure 1.3) as the radius from the axis of rotation to any given point o on the surface; r_1 as the radius from point o to the center of curvature of the meridian; and r_2 as the radius from the axis of revolution to point o, and it is perpendicular to the meridian. Then from Figure 1.3,

$$r = r_2 \sin\phi, \quad ds = r_1 \, d\phi, \quad \text{and} \quad dr = ds\cos\phi. \tag{1.1}$$

The interaction between the applied loads and resultant membrane forces is obtained from statics and is shown in Figure 1.4. Shell forces N_ϕ and N_θ are membrane forces in the meridional

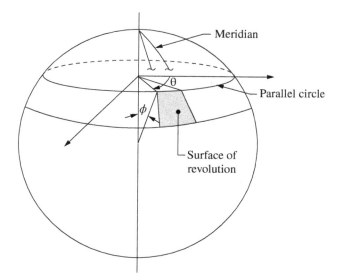

Figure 1.2   Surface of revolution

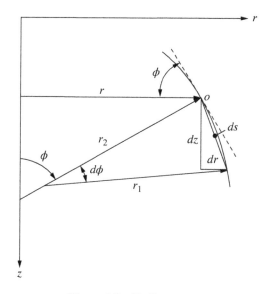

Figure 1.3   Shell geometry

and circumferential directions, respectively. Shearing forces $N_{\phi\theta}$ and $N_{\theta\phi}$ are as shown in Figure 1.4. Applied load p_r is perpendicular to the surface of the shell, load p_ϕ is in the meridional direction, and load p_θ is in the circumferential direction. All forces are positive as shown in Figure 1.4.

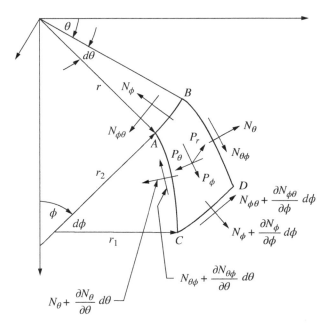

Figure 1.4   Membrane forces and applied loads

The first equation of equilibrium is obtained by summing forces parallel to the tangent at the meridian. This yields

$$N_{\theta\phi} r_1 \, d\phi - \left(N_{\theta\phi} + \frac{\partial N_{\theta\phi}}{\partial \theta} d\theta \right) r_1 \, d\phi - N_\phi r \, d\theta$$

$$+ \left(N_\phi + \frac{\partial N_\phi}{\partial \phi} d\phi \right) \left(r + \frac{\partial r}{\partial \phi} d\phi \right) d\theta \qquad (1.2)$$

$$+ p_\phi r \, d\theta r_1 \, d\phi - N_\theta r_1 \, d\phi d\theta \, \cos\phi = 0.$$

The last term in Eq. (1.2) is the component of N_θ parallel to the tangent at the meridian (Jawad 2004). It is obtained from Figure 1.5. Simplifying Eq. (1.2) and neglecting terms of higher order results in

$$\frac{\partial}{\partial \phi} \left(r N_\phi \right) - r_1 \frac{\partial N_{\theta\phi}}{\partial \theta} - r_1 N_\theta \cos\phi + p_\phi r r_1 = 0. \qquad (1.3)$$

The second equation of equilibrium is obtained from summation of forces in the direction of parallel circles. Referring to Figure 1.4,

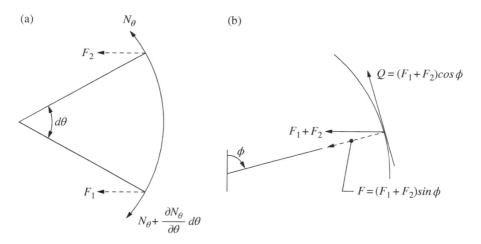

Figure 1.5 Components of N_θ: (a) circumferential cross section and (b) longitudinal cross section

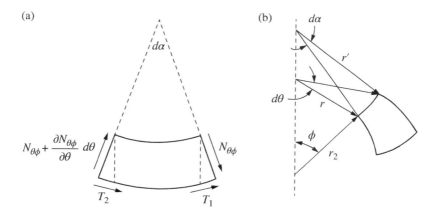

Figure 1.6 Components of $N_{\theta\phi}$: (a) side view and (b) three-dimensional view

$$N_{\phi\theta} r \, d\theta - \left(N_{\phi\theta} + \frac{\partial N_{\phi\theta}}{\partial \phi} d\phi \right) \left(r + \frac{\partial r}{\partial \phi} d\phi \right) d\theta$$

$$-N_\theta r_1 d\phi + \left(N_\theta + \frac{\partial N_\theta}{\partial \theta} d\theta \right) (r_1 d\phi)$$

$$+ p_\theta r \, d\theta r_1 d\phi - N_{\theta\phi} r_1 d\phi \frac{\cos\phi \, d\theta}{2}$$ (1.4)

$$- \left(N_{\theta\phi} + \frac{\partial N_{\theta\phi}}{\partial \theta} d\theta \right) (r_1 d\phi) \frac{\cos\phi \, d\theta}{2} = 0.$$

The last two expressions in this equation are obtained from Figure 1.6 (Jawad 2004) and are the components of $N_{\theta\phi}$ in the direction of the parallel circles. Simplifying this equation results in

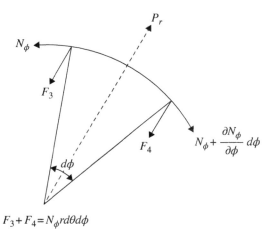

$$F_3 + F_4 = N_\phi r d\theta d\phi$$

Figure 1.7   Components of N_ϕ

$$\frac{\partial}{\partial \phi}\left(rN_{\phi\theta}\right) - r_1\frac{\partial N_\theta}{\partial \theta} + r_1 N_{\theta\phi}\cos\phi - p_\theta rr_1 = 0. \tag{1.5}$$

This is the second equation of equilibrium of the infinitesimal element shown in Figure 1.4. The last equation of equilibrium is obtained by summing forces perpendicular to the middle surface. Referring to Figures 1.4, 1.5, and 1.7,

$$(N_\theta r_1 d\phi d\theta)\sin\phi - p_r r d\theta r_1 d\phi + N_\phi r d\theta d\phi = 0$$

or

$$N_\theta r_1 \sin\phi + N_\phi r = p_r rr_1. \tag{1.6}$$

Equations (1.3), (1.5), and (1.6) are the three equations of equilibrium of a shell of revolution subjected to axisymmetric loads.

1.3   Spherical and Ellipsoidal Shells Subjected to Axisymmetric Loads

In many structural applications, loads such as deadweight, snow, and pressure are symmetric around the axis of the shell. Hence, all forces and deformations must also be symmetric around the axis. Accordingly, all loads and forces are independent of θ and all derivatives with respect to θ are zero. Equation (1.3) reduces to

$$\frac{\partial}{\partial \phi}\left(rN_\phi\right) - r_1 N_\theta \cos\phi = -p_\phi rr_1. \tag{1.7}$$

Equation (1.5) becomes

$$\frac{\partial}{\partial \phi}\left(rN_{\theta\phi}\right)+r_1 N_{\theta\phi}\cos\phi=p_\theta rr_1.$$ (1.8)

In this equation, we let the cross shears $N_{\phi\theta}=N_{\theta\phi}$ in order to maintain equilibrium. Equation (1.6) can be expressed as

$$\frac{N_\theta}{r_2}+\frac{N_\phi}{r_1}=p_r.$$ (1.9)

Equation (1.8) describes a torsion condition in the shell. This condition produces deformations around the axis of the shell. However, the deformation around the axis is zero due to axisymmetric loads. Hence, we must set $N_{\theta\phi}=p_\theta=0$ and we disregard Eq. (1.8) from further consideration.

Substituting Eq. (1.9) into Eq. (1.7) gives

$$N_\phi=\frac{1}{r_2\sin^2\phi}\left[\int r_1 r_2\left(p_r\cos\phi-p_\phi\sin\phi\right)\sin\phi d\phi+C\right].$$ (1.10)

The constant of integration C in Eq. (1.10) is additionally used to take into consideration the effect of any additional applied loads that cannot be defined by p_r and p_ϕ such as weight of contents.

Equations (1.9) and (1.10) are the two governing equations for designing double-curvature shells under membrane action.

1.3.1 Spherical Shells Subjected to Internal Pressure

For spherical shells under axisymmetric loads, the differential equations can be simplified by letting $r_1=r_2=R$. Equations (1.9) and (1.10) become

$$N_\phi+N_\theta=p_r R$$ (1.11)

and

$$N_\phi=\frac{R}{\sin^2\phi}\left[\int\left(p_r\cos\phi-p_\phi\sin\phi\right)\sin\phi d\phi+C\right].$$ (1.12)

These two expressions form the basis for developing solutions to various loading conditions in spherical shells. For any loading condition, expressions for p_r and p_ϕ are first determined and then the previous equations are solved for N_ϕ and N_θ.

For a spherical shell under internal pressure, $p_r=P$ and $p_\phi=0$. Hence, from Eqs. (1.11) and (1.12),

$$N_\phi=N_\theta=\frac{PR}{2}=\frac{PD}{4}$$ (1.13)

where D is the diameter of the sphere. The required thickness is obtained from

$$t = \frac{N_\phi}{S} = \frac{N_\theta}{S} \qquad (1.14)$$

where S is the allowable stress.

Equation (1.14) is accurate for design purposes as long as $R/t \geq 10$. If $R/t < 10$, then thick shell equations, described in Chapter 3, must be used.

1.3.2 Spherical Shells under Various Loading Conditions

The following examples illustrate the use of Eqs. (1.11) and (1.12) for determining forces in spherical segments subjected to various loading conditions.

Example 1.1

A storage tank roof with thickness t has a dead load of γ psf. Find the expressions for N_ϕ and N_θ.

Solution

From Figure 1.8a and Eq. (1.12),

$$p_r = -\gamma\cos\phi \quad \text{and} \quad p_\phi = \gamma\sin\phi$$

$$N_\phi = \frac{R}{\sin^2\phi}\left[\int\left(-\gamma\cos^2\phi - \gamma\sin^2\phi\right)\sin\phi\,d\phi + C\right]$$

$$N_\phi = \frac{R}{\sin^2\phi}\left(\gamma\cos\phi + C\right). \qquad (1)$$

As ϕ approaches zero, the denominator in Eq. (1) approaches zero. Accordingly, we must let the bracketed term in the numerator equal zero. This yields $C = -\gamma$. Equation (1) becomes

$$N_\phi = \frac{-R\gamma\left(1 - \cos\phi\right)}{\sin^2\phi}. \qquad (2)$$

The convergence of Eq. (2) as ϕ approaches zero can be checked by l'Hopital's rule. Thus,

$$N_\phi\big|_{\phi=0} = \frac{-R\gamma\sin\phi}{2\sin\phi\cos\phi}\bigg|_{\phi=0} = \frac{-\gamma R}{2}.$$

Equation (2) can be written as

$$N_\phi = \frac{-\gamma R}{1 + \cos\phi}. \qquad (3)$$

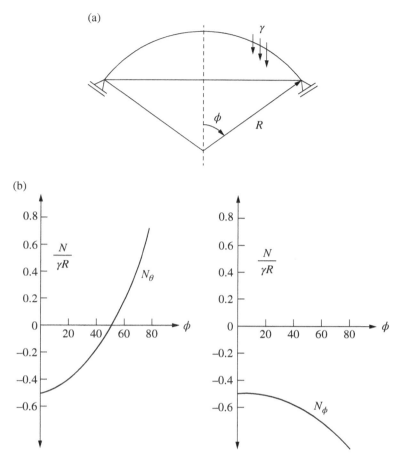

Figure 1.8 Membrane forces in a head due to deadweight: (a) dead load and (b) force patterns

From Eq. (1.11), N_θ is given by

$$N_\theta = \gamma R \left(\frac{1}{1 + \cos\phi} - \cos\phi \right). \tag{4}$$

A plot of N_ϕ and N_θ for various values of ϕ is shown in Figure 1.8b, showing that for angles ϕ greater than 52°, the hoop force, N_θ, changes from compression to tension and special attention is needed in using the appropriate allowable stress values.

Example 1.2

Find the forces in a spherical head due to a vertical load P_o applied at an angle $\phi = \phi_o$ as shown in Figure 1.9a.

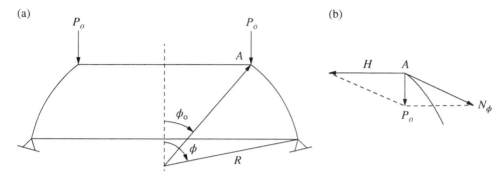

Figure 1.9   Edge loads in a spherical head: (a) edge load and (b) forces due to edge load

Solution
Since $p_r = p_\phi = 0$, Eq. (1.12) becomes

$$N_\phi = \frac{RC}{\sin^2\phi}. \tag{1}$$

From statics at $\phi = \phi_o$, we get from Figure 1.9b

$$N_\phi = \frac{P_o}{\sin\phi_o}.$$

Substituting this expression into Eq. (1), and keeping in mind that it is a compressive membrane force, gives

$$C = \frac{-P_o}{R}\sin\phi_o$$

and Eq. (1) yields

$$N_\phi = -P_o\frac{\sin\phi_o}{\sin^2\phi}.$$

From Eq. (1.11),

$$N_\theta = P_o\frac{\sin\phi_o}{\sin\phi}.$$

   In this example there is another force that requires consideration. Referring to Figure 1.9b, it is seen that in order for P_o and N_ϕ to be in equilibrium, another horizontal force, H, must be considered. The direction of H is inward in order for the force system to have a net resultant force P_o downward. This horizontal force is calculated as

$$H = \frac{-P_o\cos\phi_o}{\sin\phi_o}.$$

A compression ring is needed at the inner edge in order to contain force H. The required area, A, of the ring is given by

$$A = \frac{H(R\sin\phi_o)}{\sigma}$$

where σ is the allowable compressive stress of the ring.

Example 1.3

The sphere shown in Figure 1.10a is filled with a liquid of density γ. Hence, p_r and p_ϕ can be expressed as

$$p_r = \gamma R(1 - \cos\phi)$$
$$p_\phi = 0.$$

a. Determine the expressions for N_ϕ and N_θ throughout the sphere.
b. Plot N_ϕ and N_θ for various values of ϕ when $\phi_o = 110°$.
c. Plot N_ϕ and N_θ for various values of ϕ when $\phi_o = 130°$.
d. If $\gamma = 62.4$ pcf, $R = 30$ ft, and $\phi_o = 110°$, determine the magnitude of the unbalanced force H at the cylindrical shell junction. Design the sphere, the support cylinder, and the junction ring. Let the allowable stress in tension be 20 ksi and that in compression be 10 ksi.

(a)

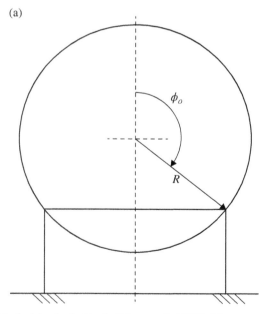

Figure 1.10   Spherical tank: (a) spherical tank, (b) support at 110°, (c) support at 130°, and (d) forces at support junction

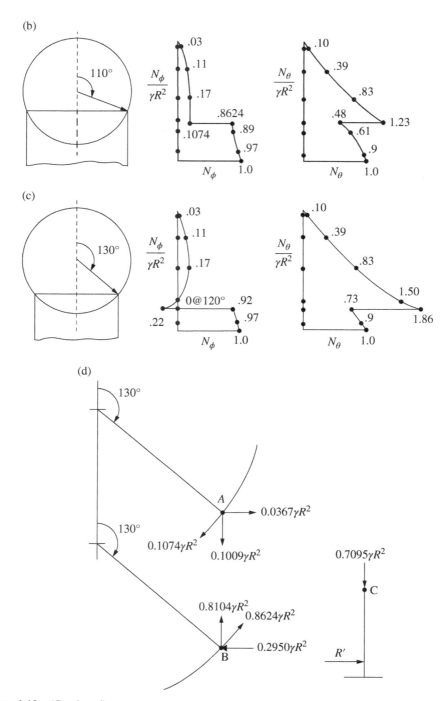

Figure 1.10   (Continued)

Solution

a. From Eq. (1.12), we obtain

$$N_\phi = \frac{\gamma R^2}{\sin^2\phi}\left(\frac{1}{2}\sin^2\phi + \frac{1}{3}\cos^3\phi + C\right). \tag{1}$$

As ϕ approaches zero, the denominator approaches zero. Hence, the bracketed term in the numerator must be set to zero. This gives $C = -1/3$ and Eq. (1) becomes

$$N_\phi = \frac{\gamma R^2}{6\sin^2\phi}\left(3\sin^2\phi + 2\cos^3\phi - 2\right). \tag{2}$$

The corresponding N_θ from Eq. (1.11) is

$$N_\theta = \gamma R^2\left[\frac{1}{2} - \cos\phi - \frac{1}{3\sin^2\phi}\left(\cos^3\phi - 1\right)\right]. \tag{3}$$

As ϕ approaches π, we need to evaluate Eq. (1) at that point to ensure a finite solution. Again the denominator approaches zero and the bracketed term in the numerator must be set to zero. This gives $C = 1/3$ and Eq. (1) becomes

$$N_\phi = \frac{\gamma R^2}{6\sin^2\phi}\left(3\sin^2\phi + 2\cos^3\phi + 2\right). \tag{4}$$

The corresponding N_θ from Eq. (1.11) is

$$N_\theta = \gamma R^2\left[\frac{1}{2} - \cos\phi - \frac{1}{3\sin^2\phi}\left(\cos^3\phi + 1\right)\right]. \tag{5}$$

Equations (2) and (3) are applicable between $0 < \phi < \phi_o$, and Eqs. (4) and (5) are applicable between $\phi_o < \phi < \pi$.

b. A plot of Eqs. (2) through (5) for $\phi_o = 110°$ is shown in Figure 1.10b. N_ϕ below circle $\phi_o = 110°$ is substantially larger than that above circle $110°$. This is due to the fact that most of the weight of the contents is supported by the spherical portion that is below the circle $\phi_o = 110°$. Also, because N_ϕ does not increase in proportion to the increase in pressure as ϕ increases, Eq. (1.11) necessitates a rapid increase in N_θ in order to maintain the relationship between the left- and right-hand sides. This is illustrated in Figure 1.10b.

A plot of N_ϕ and N_θ for $\phi_o = 130°$ is shown in Figure 1.10c. In this case, N_ϕ is in compression just above the circle $\phi_o = 130°$. This indicates that as the diameter of the supporting cylinder gets smaller, the weight of the water above circle $\phi_o = 130°$ must be supported by the sphere in compression. This results in a much larger N_θ value just above $\phi_o = 130°$. Buckling of the sphere becomes a consideration in this case.

c. From Figure 1.10b for $\phi_o = 110°$, the maximum force in the sphere is $N_\theta = 1.23\gamma R^2$. The required thickness of the sphere is

$$t = \frac{1.23(62.4)(30)^2/12}{20,000}$$

$$= 0.29 \text{ inch.}$$

A free-body diagram of the spherical and cylindrical junction at $\phi_o = 110°$ is shown in Figure 1.10d. The values of N_ϕ at points A and B are obtained from Eqs. (2) and (4), respectively. The vertical and horizontal components of these forces are shown at points A and B in Figure 1.10d. The unbalanced vertical forces result in a downward force at point C of magnitude $0.7095\gamma R^2$. The total force on the cylinder is $(0.7095\gamma R^2)(2\pi)(R)(\sin(180 - 110))$. This total force is equal to the total weight of the contents in the sphere given by (4/3) $(\pi R^3)\gamma$. The required thickness of the cylinder is

$$t = \frac{0.7095(62.4)(30)^2/12}{10,000}$$

$$= 0.33 \text{ inch.}$$

Summation of horizontal forces at points A and B results in a compressive force of magnitude $0.2583\gamma R^2$. The needed area of compression ring at the cylinder to sphere junction is

$$A = \frac{Hr}{\sigma} = \frac{0.2583 \times 62.4 \times 30^2(30\sin 70)}{10,000}$$

$$= 40.89 \text{ inch}^2.$$

This area is furnished by a large ring added to the sphere or an increase in the thickness of the sphere at the junction.

1.3.3   ASME Code Equations for Spherical Shells under Various Loading Conditions

Loading conditions such as those shown in Examples 1.1 through 1.3 are not specifically covered by equations in the boiler and pressure vessel codes. However, they are addressed in paragraph PG-16.1 of Section I, paragraph U-2(g) of Section VIII-1, and paragraph 4.1.1 of Section VIII-2 using special analysis.

In the nuclear code, paragraph NC-3932.2 of Section NC and ND-3932.2 of Section ND provide equations for calculating forces at specific locations in a shell due to loading conditions similar to those shown in Examples 1.1 through 1.5. This procedure is discussed further in Chapter 2.

1.3.4   Ellipsoidal Shells under Internal Pressure

Ellipsoidal heads of all sizes and shapes are used in the ASME code as end closure for pressure components. The general configuration is shown in Figure 1.11.

Small-size heads are formed by using dyes shaped to a true ellipse. However, large diameter heads formed from plate segments are in the shapes of spherical and torispherical geometries that simulate ellipses as shown in Figures 1.12 and 1.13. Figure 1.12 shows an ASME

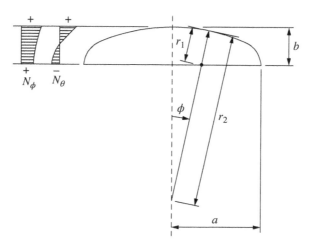

Figure 1.11   Ellipsoidal head under internal pressure

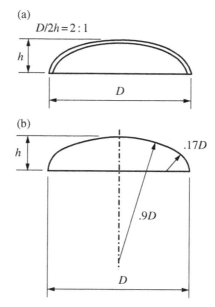

Figure 1.12   2 : 1 elliptical head: (a) exact configuration and (b) approximate configuration

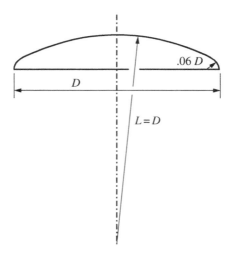

Figure 1.13   Elliptical head with $a/b = 2.96$ ratio

equivalent 2 : 1 ellipsoidal head. It consists of a spherical segment with $R = 0.9D$ and a knuckle with $r = 0.17D$ where D is the base diameter of the head. Figure 1.13 shows a shallow head (2.96 : 1 ratio) referred to as flanged and dished (F&D) head consisting of a spherical segment with $R = D$ and a knuckle section with $r = 0.06D$.

For internal pressure we define $p_r = p$ and $p_\phi = 0$. Then from Eqs. (1.1) and (1.10),

$$N_\phi = \frac{1}{r_2 \sin^2 \phi} \left(p \int r \, dr + C \right)$$

$$N_\phi = \frac{1}{r_2 \sin^2 \phi} \left(\frac{pr^2}{2} + C \right). \tag{1.15}$$

The constant C is obtained from the following boundary condition:

$$\text{At } \phi = \frac{\pi}{2}, \quad r_2 = r \quad \text{and} \quad N_\phi = \frac{pr}{2}.$$

Hence, from Eq. (1.15) we get $C = 0$ and N_ϕ can be expressed as

$$N_\phi = \frac{pr^2}{2r_2 \sin^2 \phi}$$

or

$$N_\phi = \frac{pr_2}{2}. \tag{1.16}$$

From Eq. (1.9),

$$N_\theta = pr_2 \left(1 - \frac{r_2}{2r_1}\right). \tag{1.17}$$

From analytical geometry, the relationship between the major and minor axes of an ellipse and r_1 and r_2 is given by

$$r_1 = \frac{a^2 b^2}{\left(a^2 \sin^2\phi + b^2 \cos^2\phi\right)^{3/2}}$$

$$r_2 = \frac{a^2}{\left(a^2 \sin^2\phi + b^2 \cos^2\phi\right)^{1/2}}.$$

Substituting these expressions into Eqs. (1.16) and (1.17) gives the following expressions for membrane forces in ellipsoidal shells due to internal pressure:

$$N_\phi = \frac{pa^2}{2} \frac{1}{\left(a^2 \sin^2\phi + b^2 \cos^2\phi\right)^{1/2}} \tag{1.18}$$

$$N_\theta = \frac{pa^2}{2b^2} \frac{b^2 - (a^2 - b^2)\sin^2\phi}{\left(a^2 \sin^2\phi + b^2 \cos^2\phi\right)^{1/2}}. \tag{1.19}$$

The maximum tensile force in an ellipsoidal head is at the apex as shown in Figure 1.11. The maximum value is obtained from Eqs. (1.18) and (1.19) by letting $\phi = 0$. This gives

$$N_\phi = N_\theta = \frac{Pa^2}{2b} \tag{1.20}$$

For a $2:1$ head with $a/b = 2$ and $a = D/2$, Eq. (1.18) becomes

$$N_\phi = N_\theta = 0.5PD \tag{1.21}$$

where D is the base diameter. Comparing this equation with Eq. (1.13) for spherical heads shows that the force, and thus the stress, in a $2:1$ ellipsoidal head is twice that of a spherical head having the same base diameter.

For an F&D head with $a/b = 2.96$ and $a = D/2$, Eq. (1.18) becomes

$$N_\phi = N_\theta = 0.74PD \tag{1.22}$$

Comparing this equation with Eq. (1.13) for spherical heads shows the stress of a $2.95:1$ F&D head at the apex is 2.96 times that of a spherical head having the same base diameter.

A plot of Eqs. (1.18) and (1.19) is shown in Figure 1.11. Equation (1.18) for the longitudinal force, N_ϕ, is always in tension regardless of the a/b ratio. Equation (1.19) for N_θ on the other hand gives compressive circumferential forces near the equator when the value $a/b \geq \sqrt{2}$.

For large a/b ratios under internal pressure, the compressive circumferential force tends to increase in magnitude, whereas instability may occur for large a/t ratios. This extreme care must be exercised by the engineer to avoid buckling failure. The ASME code contains design rules that take into account the instability of shallow ellipsoidal shells due to internal pressure as described in Chapter 5.

Example 1.4
Determine the required thickness of a $2:1$ ellipsoidal head with $a = 30$ inches, $b = 15$ inches, and $P = 500$ psi, and allowable stress in tension is $S = 20{,}000$ psi.

Solution
At the apex, $\phi = 0$, and from Eqs. (1.18) and (1.19),

$$N_\phi = Pa \quad \text{and} \quad N_\theta = Pa$$

At the equator, $\phi = 90°$, and from Eqs. (1.18) and (1.19),

$$N_\phi = \frac{Pa}{2} \quad \text{and} \quad N_\theta = -Pa$$

Thus, the required thickness is $t = Pa/S = 500(30)/20{,}000 = 0.75$ inch.

Notice at the equator, N_θ is compressive and may result in instability as discussed in Chapter 5.

1.4   Conical Shells

Equations (1.9) and (1.10) cannot readily be used for analyzing conical shells because the angle ϕ in a conical shell is constant. Hence, the two equations have to be modified accordingly. Referring to Figure 1.14, it can be shown that

$$\left.\begin{aligned}
\phi &= \beta = \text{constant} \\
r_1 &= \infty \quad r_2 = s\tan\alpha \quad r = s\sin\alpha \\
N_\phi &= N_s.
\end{aligned}\right] \tag{1.23}$$

Equation (1.9) can be written as

$$\frac{N_s}{r_1} + \frac{N_\theta}{s\tan\alpha} = p_r$$

or since $r_1 = \infty$,

$$\left.\begin{aligned}
N_\theta &= p_r s\tan\alpha \\
&= p_r r_2 \\
N_\theta &= \frac{p_r r}{\cos\alpha}
\end{aligned}\right] . \tag{1.24}$$

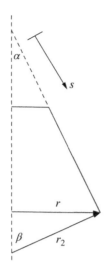

Figure 1.14 Conical shell

Similarly from Eqs. (1.1) and (1.7),

$$\frac{d}{ds}r_1(s\sin\alpha N_s)-r_1 N_\theta\sin\alpha=-p_s s\sin\alpha r_1.$$

Substituting Eq. (1.24) into this equation results in

$$N_s=\frac{-1}{s}\left[\int\int(p_s-p_r\tan\alpha)s\,ds+C\right].\qquad(1.25)$$

It is of interest to note that while N_θ is a function of N_ϕ for shells with double curvature, it is independent of N_ϕ for conical shells as shown in Eqs. (1.24) and (1.25). Also, as α approaches $0°$, Eq. (1.25) becomes

$$N_\theta=p_r r_2,$$

which is the expression for the circumferential hoop force in a cylindrical shell.

The analysis of conical shells consists of solving the forces in Eqs. (1.24) and (1.25) for any given loading condition. The thickness is then determined from the maximum forces and a given allowable stress.

Equations (1.24) and (1.25) and Figure 1.15 will be used to determine forces in a conical shell due to internal pressure.

From Eq. (1.24), the maximum N_θ occurs at the large end of the cone and is given by

$$N_\theta=p\left(\frac{r_o}{\sin\alpha}\right)\tan\alpha=\frac{p r_o}{\cos\alpha}.\qquad(1.26)$$

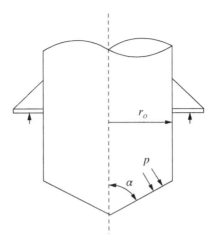

Figure 1.15   Conical bottom head

From Eq. (1.25),

$$N_s = \frac{-1}{s}\left(\int -p\tan\alpha s\,ds + C\right)$$

$$= \frac{-1}{s}\left(-p\tan\alpha\frac{s^2}{2} + C\right).$$

$$\text{At } s = L, \quad N_s = \frac{pr_o}{2}\frac{1}{\cos\alpha}$$

(1.27)

Substituting this expression into Eq. (1.27), and using the relationships of Eq. (1.23), gives $C = 0$. Equation (1.27) becomes

$$N_s = \frac{pr}{2\cos\alpha}$$

$$\text{and max } N_s = \frac{pr_o}{2\cos\alpha}.$$

(1.28)

It is of interest to note that the longitudinal and hoop forces are identical to those of a cylinder with equivalent radius of $r_o/\cos\alpha$.

All sections of the ASME code have equations for designing conical sections based on Eqs. (1.26) and (1.28).

1.5   Cylindrical Shells

Equipment consisting of cylindrical shells subjected to pressure and axial loads are frequently encountered in refineries and chemical plants. If the radius of the shell is designated by R, Figure 1.16a, then from Figure 1.3 $r_1 = \infty$, $\phi = 90°$, $P = p_r$, and $r = r_2 = R$. The value of the

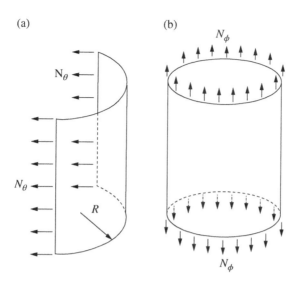

Figure 1.16 Cylindrical shell: (a) circumferential force and (b) longitudinal force

circumferential force N_θ can be obtained by equating the pressure acting on the cross section, Figure 1.16a, to the forces in the material at the cross section. This results in

$$N_\theta = p_r R \qquad (1.29)$$

The required thickness, t, of a cylindrical shell due to internal pressure is obtained from Eq. (1.29) as

$$t = \frac{PR}{S} \qquad (1.30)$$

where S is the allowable stress and t is the thickness.

The required thickness of cylindrical shells in the ASME code is obtained from a modified Eq. (1.30) that takes into consideration stress variation in the wall of the cylinder for small R/t ratios. This equation is described in Chapter 3.

Similarly, the value of the axial force N_ϕ is obtained by equating the pressure acting on the cross section, Figure 1.16b, to the forces in the material at the cross section. This yields

$$N_\phi = \frac{p_r R}{2} \qquad (1.31)$$

The corresponding stress and thickness are obtained from Eq. (1.31) as

$$t = \frac{PR}{2S} \qquad (1.32)$$

1.6   Cylindrical Shells with Elliptical Cross Section

When the cross section of a thin cylinder is elliptical, Figure 1.17, rather than circular in shape, then the relationship

$$\left(\frac{x}{a}\right)^2 + \left(\frac{y}{b}\right)^2 = 1 \tag{1.33}$$

is incorporated in the basic equations and the value of circumferential stress N_θ becomes

$$N_\theta = \frac{p_r a^2 b^2}{\left(a^2 \sin^2\phi + b^2 \cos^2\phi\right)^{3/2}} \tag{1.34}$$

where a, b, and ϕ are as defined in Figure 1.17. The value of N_ϕ is obtained from a free-body diagram at a given location.

Equation (1.34) is utilized in cases where a fabricated cylinder is slightly out-of-round and the required thickness based on the obround geometry is needed. This is illustrated in the following example.

Example 1.5
A cylindrical pressure vessel is constructed of steel plates and has an internal pressure of 100 psi. The allowable stress of steel is 20 ksi. Find the required thickness of the cylinder when the cross section is (a) circular with a diameter of 96 inches and (b) elliptical with a minor diameter of 92 inches and a major diameter of 100 inches.

Solution
a. From Eq. (1.29),

$$N_\theta = (100)\left(\frac{96}{2}\right) = 4800 \text{ lbs/inch.}$$

The thickness is obtained from the relationship $t = N_\theta/\sigma$

$$t = \frac{4800}{20,000}$$

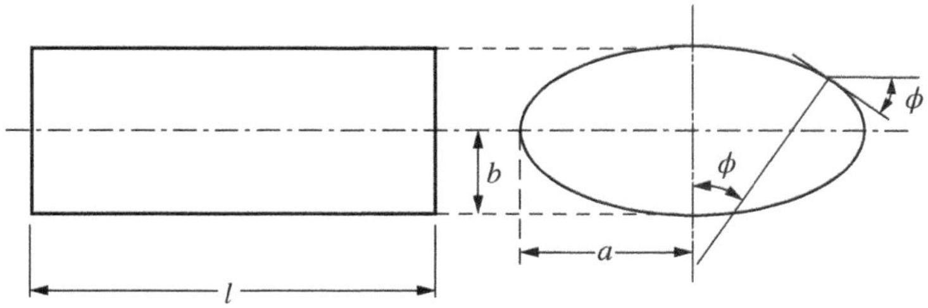

Figure 1.17   Elliptical shell

or

$$t = 0.24 \text{ inch.}$$

b. From Eq. (1.34), the maximum value of N_θ occurs at $\phi = 0°$.

$$N_\theta = \frac{(100)(50)^2(46)^2}{(50^2\sin^2 0 + 46^2\cos^2 0)^{3/2}}$$

$$N_\theta = \frac{529,000,000}{(46^2)^{3/2}}$$

$$N_\theta = 5435 \text{ lbs/inch}$$

or

$$t = \frac{5435}{20,000}$$
$$t = 0.27 \text{ inch.}$$

Hence, the thickness needs to be increased from 0.24 to 0.27 inch due to the obround shape of the cylinder.

1.7   Design of Shells of Revolution

The maximum forces for various shell geometries subjected to commonly encountered loading conditions are listed in numerous references. One such reference is by NASA (Baker et al. 1968), where extensive tables and design charts are listed. Flugge (1967) contains a thorough coverage of a wide range of applications to the membrane theory, as does Roark's handbook (Young et al. 2012).

Problems

1.1 Derive Eq. (1.10).

1.2 Determine the forces in the spherical roof of a flat bottom tank due to the snow load shown.

1.3 Determine the values of N_ϕ and N_θ in the underwater spherical shell shown. For hydrostatic pressure, let

$$p_\phi = 0$$
$$p_r = -\gamma[H + R(1 - \cos\phi)]$$

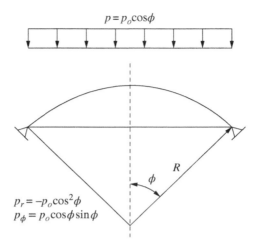

Problem 1.2   Snow load on a spherical roof

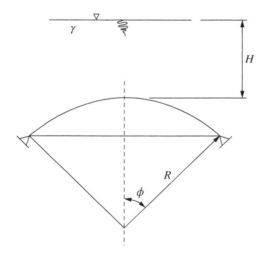

Problem 1.3   Underwater spherical shell

1.4 Determine the forces in the roof of the silo shown due to deadweight, γ. The equivalent pressure is expressed as

$$p_\phi = \gamma \sin \phi$$

$$p_r = -\gamma \cos \phi.$$

1.5 Plot the values of N_ϕ and N_θ as a function of ϕ in an ellipsoidal shell with a ratio of 3 : 1 and subjected to internal pressure.

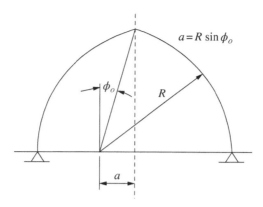

Problem 1.4 Roof of a silo

1.6 The nose of a submersible titanium vehicle is made of a 2 : 1 ellipsoidal head. Calculate the required thickness due to an external pressure of 300 psi. Let $a = 30$ inches, $b = 15$ inches, and the allowable compressive stress = 10 ksi.

1.7 An ellipsoidal head with $a/b = 2$ has a maximum stress at the apex of 20 ksi. A nozzle is required at $\phi = 45°$ (Figure 1.14). What is the stress in the head at this location in order to properly reinforce the nozzle?

1.8 Determine the maximum values and locations of N_s and N_θ in the small holding tank shown. Let $L = 22$ ft, $L_o = 2$ ft, $\gamma = 50$ pcf, and $\alpha = 30°$.

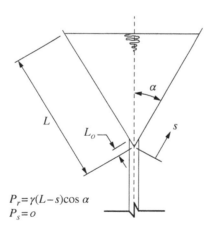

Problem 1.8 Holding tank

1.9 The lower portion of a reactor is subjected to a radial nozzle load as shown. Determine the required thickness of the conical section. Use an allowable stress of 10 ksi. What is the required area of the ring at the point of application of the load?

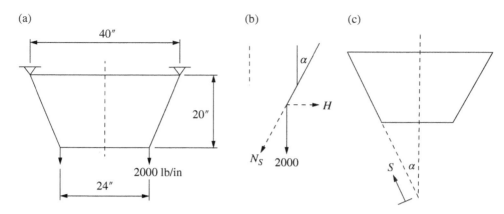

Problem 1.9 Nozzle load in a conical shell: (a) axial force, (b) components of axial force, and (c) dimension origin

2

Various Applications of the Membrane Theory

2.1 Analysis of Multicomponent Structures

The quantity $N_\phi r_2 \sin^2 \phi$ in Eq. (1.12), when multiplied by 2π, represents the total applied force acting on a structure at a given parallel circle of angle ϕ. Hence, for complicated geometries, the value of N_ϕ in Eq. (1.12) at any given location can be obtained by taking a free-body diagram of the structure. The value of N_θ at the same location can then be determined from Eq. (1.11). This method is widely used (Jawad and Farr 1989) in designing pressure vessels, flat bottom tanks, elevated water towers (Figure 2.1), and other similar structures (API 620 2014). Example 2.1 illustrates the application of this method to the design of a water-holding tank. The ASME nuclear code paragraphs NC-3932 and ND-3932 have various equations and procedures for designing components by the free-body method. This method is also useful in obtaining an approximate design at the junction of two shells of different geometries. A more accurate analysis utilizing bending moments may then be performed to establish the discontinuity stresses of the selected members at a junction if a more exact analysis is needed.

Example 2.1

The tank shown in Figure 2.2 is filled with liquid up to point a. The specific gravity is 1.0. Above point a the tank is subjected to a gas pressure of 0.5 psi. Determine the forces and thicknesses of the various components of the steel tank disregarding the deadweight of the tank. Use an allowable tensile stress of 12 000 psi and an allowable compressive stress of 8000 psi.

Solution

Tank Roof The maximum force in the roof is obtained from Figure 2.3a. Below section a–a, a 0.5 psi pressure is needed to balance the pressure above section a–a. Force N_ϕ in the roof has a

Stress in ASME Pressure Vessels, Boilers, and Nuclear Components, First Edition. Maan H. Jawad.
© 2018, The American Society of Mechanical Engineers (ASME), 2 Park Avenue,
New York, NY, 10016, USA (www.asme.org). Published 2018 by John Wiley & Sons, Inc.

Figure 2.1 Elevated water tank. Source: Courtesy of CB & I

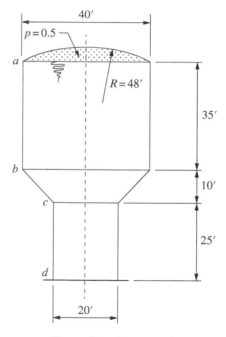

Figure 2.2 Storage tank

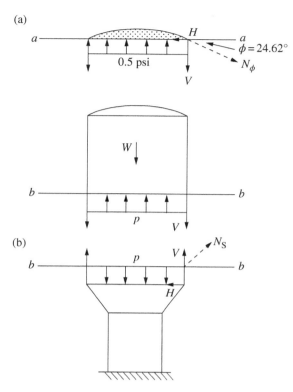

Figure 2.3   Free-body diagram of storage tank: (a) roof-to-shell junction and (b) cone-to-shell junction

vertical component V around the perimeter of the roof. Summation of forces in the vertical direction gives

$$2\pi RV - \pi R^2 P = 0$$
$$V = 60 \text{ lbs/inch.}$$

Hence,

$$N_\phi = \frac{V}{\sin\phi} = \frac{60}{0.42}$$
$$= 144 \text{ lbs/inch.}$$

From Eq. (1.11) with $r_1 = r_2 = R$,

$$N_\theta = 144 \text{ lbs/inch.}$$

Since N_θ and N_ϕ are the same, use either one to calculate the thickness. The required thickness is

$$t = \frac{N_\phi}{\sigma} = \frac{144}{12{,}000}$$
$$= 0.012 \text{ inch.}$$

Because this thickness is impractical to handle during fabrication of a tank with such a diameter, use $t = 1/4$ inch.

40-Ft Shell The maximum force in the shell is at section b–b as shown in Figure 2.3b. The total weight of liquid at section b–b is

$$W = 62.4(\pi)(20)^2(35)$$

$$= 2,744,500 \text{ lbs.}$$

Total pressure at b–b is

$$P = 0.5 + \left(\frac{62.4}{144}\right)(35)$$

$$P = 15.67 \text{ psi.}$$

The total sum of the vertical forces at b–b is equal to zero. Hence,

$$2,744,500 - (15.67)(\pi)(240)^2 + V(\pi)(480) = 0$$

or

$$V = 60 \text{ lbs/inch}$$

and

$$N_\phi = 60 \text{ lbs/inch.}$$

In a cylindrical shell, $r_1 = \infty$ and $r_2 = R$. Hence, Eq. (1.11) becomes

$$N_\theta = pR = (15.67)(240) = 3761 \text{ lbs/inch.}$$

The required thickness

$$t = \frac{N_\theta}{\sigma} = \frac{3761}{12,000}.$$

$$t = 0.031 \text{ inch.}$$

Use $t = 3/8$ inch in order to better match the conical transition section discussed later. The unbalanced force at the roof-to-shell junction is

$$H = 131 \text{ lbs/inch (inwards).}$$

The area required to contain the unbalanced force, H, is given by

$$A = (H)\frac{\text{shell radius}}{\text{allowable compressive stress}}$$

$$= 131 \times \frac{240}{8000} = 3.93 \text{ inch}^2.$$

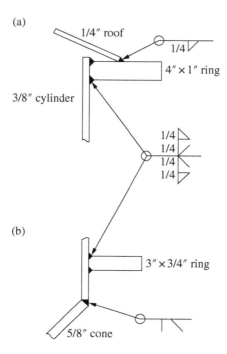

Figure 2.4 Roof-to-large cylinder and large cylinder-to-cone junctions: (a) roof-to-shell stiffener and (b) cone-to-shell stiffener

Use 1 inch thick × 4 inch wide ring as shown in Figure 2.4a.

Conical Transition At section *b–b*, force *V* in the 40-ft shell must equal force *V* in the cone in order to maintain equilibrium as shown in Figure 2.3b. Thus,

$$V = 60 \text{ lbs/inch.}$$

and

$$N_s = \frac{60}{0.707} = 85 \text{ lbs/inch.}$$

In a conical shell $r_1 = \infty$ and $r_2 = R/\cos \alpha$. Hence from Eq. (1.30),

$$N_\theta = \frac{pR}{\cos \alpha} = \frac{240(15.67)}{0.707} = 5319 \text{ lbs/inch.}$$

The required thickness at the large end of the cone is

$$t = \frac{5319}{12,000} = 0.44 \text{ inch.}$$

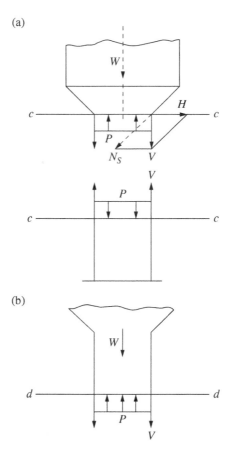

Figure 2.5 Cone–cylinder junction: (a) forces at small end of cone and (b) forces in small cylinder

The horizontal force at point b is $N_s \cos 45$.

$$H = 60 \text{ lbs/inch (inwards)}$$

The required area is

$$A = 60 \times \frac{240}{8000} = 1.8 \text{ inch}^2.$$

Use 3/4 inch thick by 3 inch wide ring as shown in Figure 2.4b.

The forces at the small end of the cone are shown in Figure 2.5a. The weight of the liquid in the conical section at point c is

$$W = \frac{\pi \gamma H \left(r_1^2 + r_1 r_2 + r_2^2 \right)}{3}$$

$$= \frac{\pi \times 62.4 \times 10 \left(10^2 + 10 \times 20 + 20^2 \right)}{3}$$

$$= 457,400 \text{ lbs.}$$

Total liquid weight is

$$W = 2,744,500 + 457,400$$

$$= 3,201,900 \text{ lbs.}$$

Pressure at section c–c is

$$P = 0.5 + \left(\frac{62.4}{144}\right)(45)$$

$$= 20.0 \text{ psi.}$$

Summing forces at section c–c gives

$$20.0 \times \pi \times 120^2 - 3,201,900 - (V \times \pi \times 240) = 0$$

$$V = -3047 \text{ lbs/inch.}$$

The negative sign indicates that the vertical component of N_s is opposite to that assumed in Figure 2.5a and is in compression rather than tension. This is caused by the column of liquid above the cone whose weight is greater than the net pressure force at section c–c.

$$N_s = \frac{-3047}{0.707}$$

$$= -4309 \text{ lbs/inch (compressive)}$$

$$N_\theta = \frac{pR}{\cos\alpha} = \frac{20.0 \times 120}{0.707}$$

$$= 3395 \text{ lbs/inch.}$$

N_θ at the small end is smaller than N_θ at the large end. Hence, the thickness at the small end need not be calculated for N_θ. Since N_s at the small end is in compression, the thickness due to this force needs to be calculated because the allowable stress in compression is smaller than that in tension. Hence,

$$t = \frac{4309}{8000} = 0.54 \text{ inch.}$$

Use $t = 5/8$ inch for the cone.

The horizontal force at section c–c, Figure 2.5, is given by

$$H = 3047 \text{ lbs/inch inwards.}$$

The required area of the ring is

$$A = \frac{3047 \times 120}{8000} = 45.71 \text{ inch}^2.$$

This large required area is normally distributed around the junction as shown in Figure 2.6.

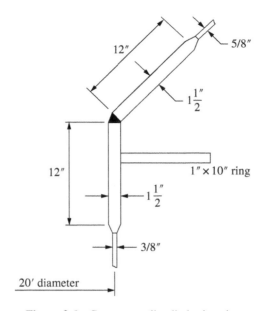

Figure 2.6   Cone-to-small cylinder junction

20-Ft Shell   At section c–c, Figure 2.5, the value of V in the 20-ft shell is the same as V in the cone due to continuity. Thus,

$$N_\phi = V = -3047 \text{ lbs/inch}$$
$$N_\theta = pR = 20.0 \times 120 = 2400 \text{ lbs/inch}.$$

At section d–d, Figure 2.5, the liquid weight is given by

$$W = 3,201,900 + (62.4)(\pi)(10)^2(25)$$
$$= 3,692,000 \text{ lbs}$$

and the pressure is calculated as

$$p = 0.5 + \left(\frac{62.4}{144}\right)(70)$$
$$= 30.83 \text{ psi}.$$

From Figure 2.5b, the summation of forces about d–d gives

$$3,692,000 - 30.83 \times \pi \times 120^2 + V \times \pi \times 240 = 0$$

or

$$N_\phi = V = -3047 \text{ lbs/inch},$$

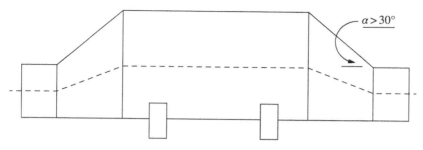

Figure 2.7 Kettle boiler

which is the same as that at point *c*.

$$N_\theta = pR = 30.83 \times 120$$

$$= 3700 \text{ lbs/inch.}$$

The required thickness of the shell is governed by N_ϕ at section *d–d*.

$$t = \frac{3047}{8000} = 0.38 \text{ inch.}$$

Use $t = 3/8$ inch for bottom cylindrical shell.

The procedure described above is sometimes used to evaluate components of equipment designed in accordance with VIII-1 that are outside the scope of the given rules. These include kettle boilers, Figure 2.7, with conical transition sections with one-half apex angle greater than the 30° limit set in VIII-1. Another example is complex geometries such as the one shown in Figure 2.8 for a cone with a double-axis cross section.

2.2 Pressure–Area Method of Analysis

The membrane theory is very convenient in determining thicknesses of major components such as cylindrical, conical, hemispherical, and ellipsoidal shells. The theory, however, is inadequate for analyzing complicated geometries such as nozzle attachments, transition sections, and other details similar to those shown in Figure 2.9. An approximate analysis of these components can be obtained by using the pressure–area method. A more accurate analysis can then be performed based on the bending theories of Chapters 3 and 4 or the finite element theory.

The pressure–area analysis is based on the concept (Zick and St. Germain 1963) that the pressure contained in a given area within a shell must be resisted by the metal close to that area. Referring to Figure 2.10a, the total force in the shaded area of the cylinder is $(r)(P)(L)$, while the force supported by the available metal is $(L)(t)(\sigma)$. Equating these two expressions results in

Figure 2.8   Double-cone vessel (Nooter Corp, Saint Louis, MO)

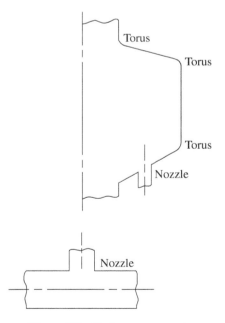

Figure 2.9   Various components

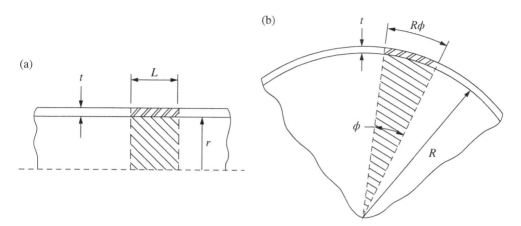

Figure 2.10   Pressure–area interaction: (a) cylindrical shell and (b) spherical shell

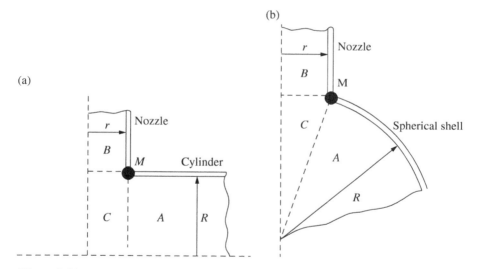

Figure 2.11   Nozzle junctions: (a) nozzle-to-shell junction and (b) nozzle-to-head junction

$t = Pr/\sigma$, which is the equation for the required thickness of a cylindrical shell. Similarly for spherical shells, Figure 2.10b gives

$$(R\phi)(R)(P)\left(\frac{1}{2}\right) = (R\phi)(t)(\sigma)$$

$$t = \frac{PR}{2\sigma}.$$

Referring to Figure 2.11a, it is seen that pressure–area A is contained by the cylinder wall and pressure–area B is contained by the nozzle wall. However, pressure–area C is not contained by any material. Thus we must add material, M, at the junction. The area of material M is given by

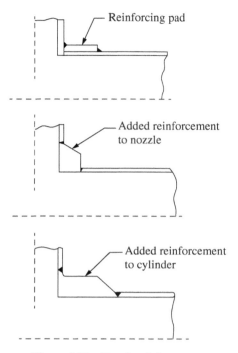

Figure 2.12   Nozzle reinforcement

$$(P)(R)(r) = (\sigma)(M)$$

$$M = \frac{P(R)(r)}{\sigma}.$$

For a spherical shell, the required area, M, from Figure 2.11b is

$$(P)(R)(r)\left(\frac{1}{2}\right) = (\sigma)(M)$$

$$M = \left(\frac{1}{2}\right)\frac{(P)(R)(r)}{\sigma}.$$

The required area is added either to the shell, nozzle, or as a reinforcing pad as shown in Figure 2.12.

The pressure–area method can also be applied for junctions Farr and Jawad (2010) between components as shown in Figure 2.13. Referring to Figure 2.13b, the spherical shell must contain the pressure within area ABC. The cylindrical shell contains the pressure within area $AOCD$. At point A where the spherical and cylindrical shells intersect, the pressure–area to be contained at point A is given by AOC. However, because area AOC is used both in the ABC area for the sphere and $AOCD$ for the cylinder, and because it can be used only once, this

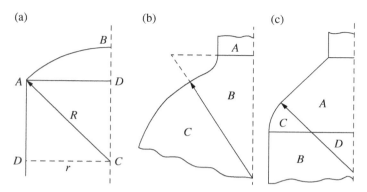

Figure 2.13 Various shell junctions: (a) head-to-shell junction, (b) flue-to-shell junction, and (c) knuckle-to-shell junction

area must be subtracted from the total calculated pressure in order to maintain equilibrium. In other words, this area causes compressive stress at point A. The area required is given by

$$A = \frac{(r)\left(\sqrt{R^2 - r^2}\right)(1/2)(P)}{\sigma}$$

where σ is allowed compressive stress.

In Figure 2.13b, pressure–area A is contained by the cylindrical shell and area C by the spherical shell. Area B is contained by the transition shell, which is in tension because area B is used neither in the area A nor area C calculations.

In Figure 2.13c, pressure–area A is contained by the cone and area B by the cylinder. The transition shell between the cone and the cylinder contains pressure–area C that is in tension and area D that is in compression. Summation of areas C and D will determine the state of stress in the transition shell.

Example 2.2
Find the required thickness of the cylindrical, spherical, and transition shells shown in Figure 2.14. Let $p = 150$ psi and $\sigma = 15,000$ psi.

Solution
For the cylindrical shell,

$$t = \frac{pr}{\sigma} = \frac{150 \times 36}{15,000} = 0.36 \text{ inch.}$$

For the spherical shell,

$$t = \frac{pR}{2\sigma} = \frac{150 \times 76}{2 \times 15,000} = 0.38 \text{ inch.}$$

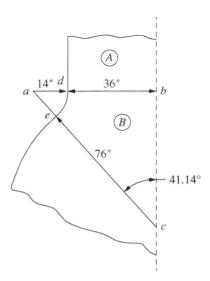

Figure 2.14   Cylindrical-to-spherical transition area

For the transition shell, pressure–area B = area of triangle abc – area of segment ade

$$= 150\left[\left(76\cos 41.14 \times \frac{50}{2}\right) - 14^2\left(48.86 \times \frac{\pi}{180}\right)/2\right]$$

$$= 202,110 \text{ lbs.}$$

$$t \times \sigma \times 14\left(\frac{48.86\pi}{180}\right) = 202,110$$

or

$$t = 1.13 \text{ inch.}$$

The pressure–area method becomes handy in situations where the ASME code does not have equations to cover particular details. One such example is shown in Figure 2.15 for a double spherical vessel where the reinforcement of the junction between the spheres can easily be obtained from the pressure–area method. Another example is shown in Figure 2.16 for a flue between a conical transition section and a cylindrical shell. The thickness of the flue can be calculated from (Farr and Jawad 2010)

$$t_f = \frac{180}{\alpha\pi r}\left[\frac{P(C_1 + C_2 + C_3)}{1.5SE} - C_4 - C_5\right] \tag{2.1}$$

where
α = flue angle, degree (the flue angle is normally the same as the cone angle)
$C_1 = 0.125(2r + D_1)^2\tan\alpha - \alpha\pi r^2/360$
$C_2 = 0.28D_1(D_1 t_s)^{0.5}$
$C_3 = 0.78C_6(C_6 t_c)^{0.5}$
$C_4 = 0.78t_c(C_6 t_c)^{0.5}$

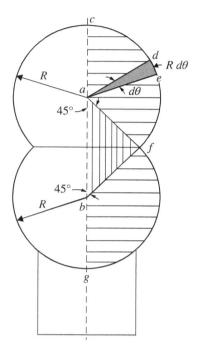

Figure 2.15   Spherical vessel

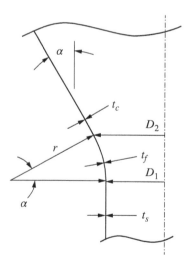

Figure 2.16   Flue transition between conical and cylindrical sections

$C_5 = 0.55 t_s (D_1 t_s)^{0.5}$

$C_6 = [D_1 + 2r(1 - \cos \alpha)]/(2\cos \alpha)$

E = joint efficiency

P = internal pressure

r = radius of flue

S = allowable stress

t_c = thickness of cone

t_f = thickness of flue

t_s = thickness of shell

The pressure–area method is also commonly used to design fittings and other piping components. Figure 2.17 shows one design method for some components.

Another application of the pressure–area method is in analyzing conduit bifurcations, Figure 2.18. The analysis of such structures is discussed by Swanson et al. (1955) and AISI (1981).

2.3   Deflection Due to Axisymmetric Loads

The deflection of a shell due to membrane forces caused by axisymmetric loads can be derived from Figure 2.19. The change of length AB due to deformation is given by

$$\frac{dv}{d\phi} d\phi - w \, d\phi.$$

The strain is obtained by dividing this expression by the original length $r_1 \, d\phi$

$$\varepsilon_\phi = \frac{1}{r_1} \frac{dv}{d\phi} - \frac{w}{r_1}. \tag{2.2}$$

The increase in radius r due to deformation, Figure 2.20, is given by

$$v \cos \phi - w \sin \phi$$

or

$$\varepsilon_\theta = \frac{1}{r}(v \cos \phi - w \sin \phi).$$

Substituting in this equation the value

$$r = r_2 \sin \phi$$

gives

$$\varepsilon_\theta = \frac{v}{r_2} \cot \phi - \frac{w}{r_2} \tag{2.3}$$

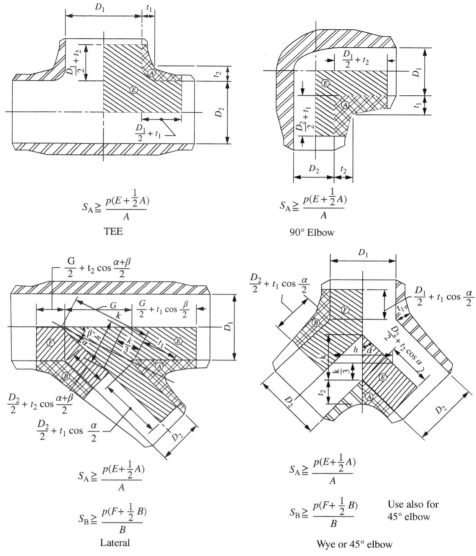

$$S_A \geqq \frac{p(E + \frac{1}{2}A)}{A}$$

TEE

$$S_A \geqq \frac{p(E + \frac{1}{2}A)}{A}$$

90° Elbow

$$S_A \geqq \frac{p(E + \frac{1}{2}A)}{A}$$

$$S_B \geqq \frac{p(F + \frac{1}{2}B)}{B}$$

Lateral

$$S_A \geqq \frac{p(E + \frac{1}{2}A)}{A}$$

$$S_B \geqq \frac{p(F + \frac{1}{2}B)}{B}$$ Use also for 45° elbow

Wye or 45° elbow

Nomenclature

A, B	- Metal area (inch^2)	
D_1, D_2	- Inside diameter of fittings (inch)	
E, F	- Indicated pressure area (inch^2)	
G, h, k	- Indicated lengths (inch)	
P	- Design pressure at design temperature (psi)	

S_A, S_B - Allowable stress at design temperature (psi)
t_1, t_2 - Indicated metal thickness (inch)
t_3 - Average metal thickness of flat surface (inch)
α, β - Indicated angles

Figure 2.17 Fitting reinforcement. Source: The M. W. Kellogg Company 1961. Reproduced with permission of John Wiley & Sons, Inc.

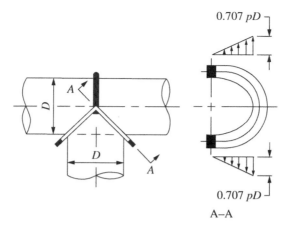

Figure 2.18 Conduit reinforcement. Source: Courtesy of Steel Plate Fabricators Association

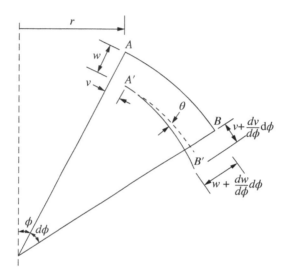

Figure 2.19 Deflection of shell

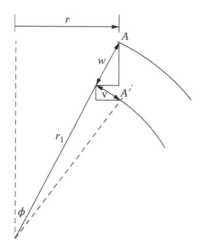

Figure 2.20 Radius increment due to deflection

or

$$w = v\cot\phi - r_2\varepsilon_\theta. \tag{2.4}$$

From Eqs. (2.2) and (2.3), we get

$$\frac{dv}{d\phi} - v\cot\phi = r_1\varepsilon_\phi - r_1\varepsilon_\theta. \tag{2.5}$$

From theory of elasticity for two-dimensional problems,

$$\left.\begin{aligned}
\varepsilon_\phi &= \frac{1}{Et}\left(N_\phi - \mu N_\theta\right)\\
\varepsilon_\theta &= \frac{1}{Et}\left(N_\theta - \mu N_\phi\right).
\end{aligned}\right] \tag{2.6}$$

Hence, Eq. (2.5) becomes

$$\frac{dv}{d\phi} - v\cot\phi = \frac{1}{Et}\left[N_\phi(r_1 + \mu r_2) - N_\theta(r_1 + \mu r_2)\right] \tag{2.7}$$

Let the right-hand side of this equation be expressed as $g(\phi)$,

$$g(\phi) = \frac{1}{Et}\left[N_\phi(r_1 + \mu r_2) - N_\theta(r_1 + \mu r_2)\right] \tag{2.8}$$

a solution of Eq. (2.7) is

$$v = \sin\phi\left(\int \frac{g(\phi)}{\sin\phi} + C\right). \tag{2.9}$$

In order to solve for the deflections in a structure due to a given loading condition, we first obtain N_ϕ and N_θ from Eqs. (1.12) and (1.11). We then calculate v from Eqs. (2.8) and (2.9). The normal deflection, w, is then calculated from Eq. (2.4).

The rotation at any point is obtained from Figures 2.19 and 2.20 as

$$\psi = \frac{1}{r_1}\left(\frac{dw}{d\phi} + v\right). \tag{2.10}$$

Example 2.3

Find the deflection at point A of the dome roof shown in Figure 2.21 due to an internal pressure p. Let $\mu = 0.3$.

Solution

For a spherical roof with internal pressure,

$$r_1 = r_2 = R$$

$$p_r = p \quad \text{and} \quad p_\phi = 0.$$

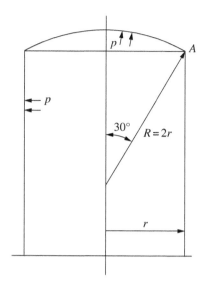

Figure 2.21   Flat bottom tank

The membrane forces are obtained from Eqs. (1.12) and (1.11) as

$$N_\phi = N_\theta = \frac{pR}{2}.$$

From Eq. (2.8),

$$g(\phi) = \frac{1}{Et}\left[\left(\frac{pR}{2}\right)(R)(1+\mu) - \left(\frac{pR}{2}\right)(R)(1+\mu)\right]$$

$$= 0$$

and Eq. (2.9) gives

$$v = C\sin\phi$$

at the point of support, the deflection $v = 0$. Hence, $C = 0$.
   From Eq. (2.4)

$$w = -r_2\varepsilon_\theta$$

and from Eq. (2.6),

$$w = \frac{-pR^2}{2Et}(1-\mu). \tag{2.11}$$

At point A, the horizontal component of the deflection is given by

$$w_h = \frac{pR^2}{2Et}(1-\mu)\sin\phi.$$

Hence, with $\mu = 0.3$ and $\phi = 30°$

$$w_h = \frac{0.175pR^2}{Et}.$$

If we substitute into this expression the quantity $R = 2r$, which is obtained from Figure 2.21, we get

$$w_h = \frac{0.7pr^2}{Et}. \tag{1}$$

It will be seen in the next chapter that the deflection of a cylinder due to internal pressure is given by

$$w = \frac{pr^2}{Et}\left(1 - \frac{\mu}{2}\right)$$

or, for $\mu = 0.3$,

$$w = \frac{0.85pr^2}{Et} \tag{2}$$

A comparison of Eqs. (1) and (2) shows that at the roof-to-cylinder junction, there is an offset in the calculated horizontal deflection. However, this offset does not exist in a real structure because of discontinuity forces that normally develop at the junction. These forces consist of local bending moments and shear forces. Although these forces eliminate the deflection offset, they do create high localized bending stresses. It turns out that these localized stresses are secondary in nature and can be discarded in the design of most structures as described in the next chapter.

Problems

2.1 Determine the thickness of all components including stiffener rings at points A and B of the steel tower. The tower is full of water to point A. The roof is subjected to a snow load of 25 psf. The allowable stress in tension is 15 ksi and the allowable stress in compression is 300 psi for the roof and 10 ksi for other components.

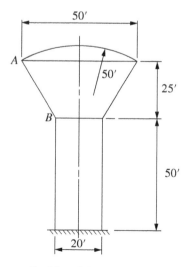

Problem 2.1   Steel tower

2.2 Determine the thickness of all components including stiffener rings of the elevated water
     tower. The tower is full of liquid between points A and D. The allowable stress in tension is
     15 ksi and that in compression is 10 ksi.

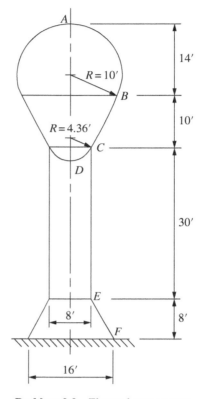

Problem 2.2   Elevated water tower

2.3 Using the pressure–area method, show that the required thickness of the bellows expansion
     joint due to internal pressure can be approximated by

$$t = \frac{p(d+w)}{s(1.14 + 4w/q)}$$

where
$d \gg w$
s = allowable stress
p = internal pressure

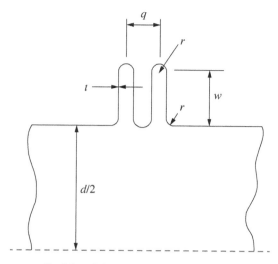

Problem 2.3   Bellows expansion joint

2.4 Calculate the area needed to reinforce the nozzle shown. Let $p = 800$ psi and $\sigma = 20$ ksi.

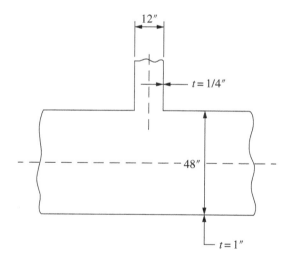

Problem 2.4   Nozzle in a shell

2.5 Calculate the thicknesses t_1, t_2, t_3, t_4, and area A of the section shown based on an average allowable stress of 18 ksi. Let $p = 250$ psi.

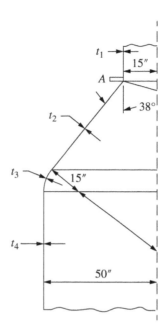

Problem 2.5   Portion of a pressure vessel

2.6 The pipe elbow is subjected to a pressure of 4000 psi. Show that the average stress in the outer surface is equal to 10,170 psi and the stress in the inner surface is equal to 28,230 psi.

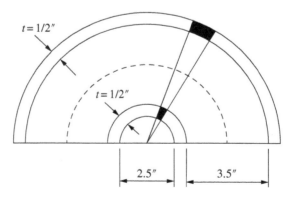

Problem 2.6   Cross section of a pipe elbow

3

Analysis of Cylindrical Shells

3.1 Elastic Analysis of Thick-Wall Cylinders

The membrane stress calculated from Eq. (1.27) tends to underestimate the actual stress in a heavy wall cylinder, Figure 3.1, with an inside radius to thickness ratio of less than 10. This is due to the nonlinear stress distribution that develops across the wall of a thick cylinder when an internal pressure is applied.

In this chapter the equations for a heavy wall cylinder are derived for elastic, plastic, and creep conditions. We start with the elastic solution. The stress equations across the wall of a cylinder when R_i/t is less than 10 are obtained from Figure 3.2. Summation of forces in the r direction gives

$$(\sigma_r + d\sigma_r)(r + dr)d\theta - \sigma_r r d\theta - 2\sigma_\theta dr \sin\left(\frac{d\theta}{2}\right) = 0.$$

Simplifying terms, deleting values of higher order, and approximating $\sin(d\theta/2)$ by $(d\theta/2)$ result in

$$\sigma_\theta - \sigma_r - r d\sigma_r/dr = 0. \tag{3.1}$$

In addition to the force equilibrium Eq. (3.1), equations for stress–strain relationships are needed. These are obtained from the theory of elasticity as

$$\sigma_r = \frac{E}{1-\mu^2}(\varepsilon_r + \mu\varepsilon_\theta) \tag{3.2}$$

Stress in ASME Pressure Vessels, Boilers, and Nuclear Components, First Edition. Maan H. Jawad.
© 2018, The American Society of Mechanical Engineers (ASME), 2 Park Avenue,
New York, NY, 10016, USA (www.asme.org). Published 2018 by John Wiley & Sons, Inc.

Figure 3.1 Thick wall vessel during fabrication. Source: Courtesy of Nooter Corporation, St. Louis

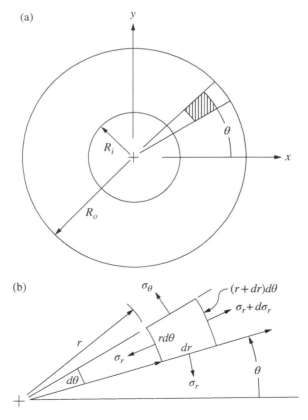

Figure 3.2 Thick-walled cylinder: (a) cross section and (b) infinitesimal section

and

$$\sigma_\theta = \frac{E}{1-\mu^2}(\varepsilon_\theta + \mu\varepsilon_r) \tag{3.3}$$

where

E = modulus of elasticity
ε_r = radial strain
ε_θ = circumferential strain
μ = Poisson's ratio
σ_r = radial stress
σ_θ = circumferential stress

The relationship between strain and deformation is obtained from Figure 3.2:

$$\varepsilon_r = \frac{dw}{dr} \tag{3.4}$$

and

$$\varepsilon_\theta = \frac{w}{r} \tag{3.5}$$

where w is radial deflection. Substituting Eqs. (3.2) through (3.5) into Eq. (3.1) results in

$$\frac{d^2w}{dr^2} + \frac{1}{r}\frac{dw}{dr} - \frac{w}{r^2} = 0. \tag{3.6}$$

One solution of this differential equation is in the form

$$w = K_1 r + K_2/r. \tag{3.7}$$

Substituting Eqs. (3.5) through (3.7) into Eqs. (3.2) and (3.3) gives

$$\sigma_r = \frac{E}{1-\mu^2}\left[K_1(1+\mu) - K_2(1-\mu)/r^2\right] \tag{3.8}$$

and

$$\sigma_\theta = \frac{E}{1-\mu^2}\left[K_1(1+\mu) + K_2(1-\mu)/r^2\right]. \tag{3.9}$$

The constants K_1 and K_2 are obtained from Eq. (3.8) by letting $\sigma_r = -p_i$ at $r = R_i$ and $\sigma_r = -p_o$ at $r = R_o$. This gives

$$K_1 = \frac{1-\mu}{E}\frac{R_i^2 p_i - R_o^2 p_o}{R_o^2 - R_i^2} \tag{3.10}$$

$$K_2 = \frac{1+\mu}{E}\frac{R_i^2 R_o^2(p_i - p_o)}{R_o^2 - R_i^2}. \tag{3.11}$$

With K_1 and K_2 known, stresses σ_r and σ_θ are obtained from Eqs. (3.8) and (3.9) as

$$\sigma_r = \frac{p_i R_i^2 - p_o R_o^2 - (p_i - p_o)\left(R_i^2 R_o^2 / r^2\right)}{R_o^2 - R_i^2} \tag{3.12}$$

$$\sigma_\theta = \frac{p_i R_i^2 - p_o R_o^2 + (p_i - p_o)\left(R_i^2 R_o^2 / r^2\right)}{R_o^2 - R_i^2}. \tag{3.13}$$

The expression for axial stress, σ_L, is given as follows:

For cylinders with open ends,

$$\sigma_L = 0 \tag{3.14}$$

For cylinders with closed ends,

$$\sigma_L = \frac{p_i R_i^2 - p_o R_o^2}{R_o^2 - R_i^2} \tag{3.15}$$

where

$r =$ radius at any point in the shell
$R_i =$ inside radius
$R_o =$ outside radius
$p_i =$ inside pressure
$p_o =$ outside pressure

The expression for the deflection of a cylinder due to internal and external pressures is obtained by substituting Eqs. (3.10) and (3.11) into Eq. (3.7):

$$w = \frac{1}{E\left(R_o^2 - R_i^2\right)}\left[p_i\left(R_i^2 r + \frac{R_i^2 R_o^2}{r}\right) - \mu p_i\left(R_i^2 r - \frac{R_i^2 R_o^2}{r}\right) - p_o\left(R_o^2 r + \frac{R_i^2 R_o^2}{r}\right).\right.$$
$$\left. + \mu p_o\left(R_o^2 r - \frac{R_i^2 R_o^2}{r}\right)\right] \tag{3.16}$$

The deflection at the inside surface, w_i, is obtained by letting $r = R_i$, and the deflection at the outer surface, w_o, is obtained by letting $r = R_o$:

$$w_i = \frac{1}{E\left(R_o^2 - R_i^2\right)}\left[R_i p_i\left(R_o^2 + R_i^2\right) + \mu R_i p_i\left(R_o^2 - R_i^2\right) - 2p_o R_i R_o^2\right] \tag{3.17}$$

$$w_o = \frac{1}{E\left(R_o^2 - R_i^2\right)}\left[2p_i R_i^2 R_o - R_o p_o\left(R_o^2 + R_i^2\right) + \mu R_o p_o\left(R_o^2 - R_i^2\right)\right]. \tag{3.18}$$

For thin cylindrical shells, the deflection given by Eqs. (3.17) and (3.18) can be approximated by the equation

$$w = \frac{PR^2}{Et} \qquad (3.19)$$

where P is the applied internal pressure, R is the inside radius, and t is the thickness.

Equations (3.12) through (3.15) are referred to as Lame's equations. The expression for the circumferential stress in the shell due to internal pressure p_i with zero external pressure p_o is obtained from Eq. (3.13) as

$$\sigma_\theta = \frac{p_i R_i^2 \left(1 + R_o^2/r^2\right)}{R_o^2 - R_i^2}. \qquad (3.20)$$

The maximum circumferential stress is at the inside surface, $r = R_i$, is

$$\sigma_\theta = \frac{p_i \left(R_o^2 + R_i^2\right)}{R_o^2 - R_i^2} \qquad (3.21a)$$

$$\sigma_\theta = \frac{p_i(\gamma^2 + 1)}{\gamma^2 - 1} \qquad (3.21b)$$

where $\gamma = R_o/R_i$. Figure 3.3 shows this stress distribution and indicates a maximum stress on the inside surface of the cylinder. A comparison of the maximum stress from this equation and from Eq. (1.36) for thin shells shows a small difference of about 5% in stress magnitude when the

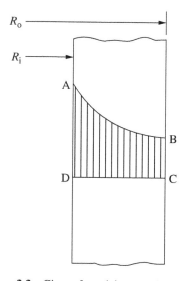

Figure 3.3 Circumferential stress distribution

ratio of $R_i/t \approx 10$. However, this difference becomes significant for lower R_i/t ratios. The required thickness, t, is obtained from Eq. (3.21) by letting $R_o = R_i + t$:

$$t = R_i\left(Z^{0.5} - 1\right) \tag{3.22}$$

where

p_i = internal pressure
R_i = inside radius
$S = \sigma_\theta$ = allowable stress
t = thickness
$Z = (S + p_i)/(S - p_i)$

3.2   Thick Cylinders with Off-center Bore

The maximum elastic circumferential stress in a thick cylinder with an off-center bore occurs at the thin section, point A in Figure 3.4. The governing equation is expressed as (Faupel and Fisher 1981)

$$\sigma_\theta = P\left[\frac{2\,R_o^2\left(R_o^2 + R_i^2 - 2\rho R_i - \rho^2\right)}{\left(R_o^2 + R_i^2\right)\left(R_o^2 - R_i^2 - 2\rho R_i - \rho^2\right)} - 1\right]. \tag{3.23}$$

It is of interest to note that Eq. (3.23) gives a lower stress value than Eq. (3.21) using an equivalent cylinder with an inside radius, R_i, and a pseudo outside radius, $R_o' = R_o - \rho$, due to the stiffness of the material adjacent to the thin section.

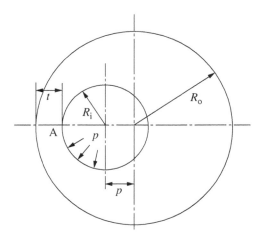

Figure 3.4   Cylinder with eccentric bore

3.3   Stress Categories and Equivalent Stress Limits for Design and Operating Conditions

The stress distribution in Figure 3.3 due to internal pressure must be decomposed into various stress categories in order to establish stress limits in accordance with Sections VIII-2, III-NC, and III-ND. This process is described next. For this purpose the ASME code defines the following stress categories present at a given cross section:

1. Primary stress. Primary stress is any normal or shear stress developed by an imposed loading, which is necessary to satisfy the laws of equilibrium of external and internal forces and moments. The basic characteristic of a primary stress is that it is not self-limiting. Primary stress is divided into three subcategories: general membrane stress, local membrane stress, and bending stress:
   a. General membrane stress, P'_m. This stress is so distributed in the structure that no redistribution of load occurs as a result of yielding. An example is the stress in a cylindrical or spherical shell, away from discontinuities, due to internal pressure or to distributed live load.
   b. Local membrane stress, P'_L. Cases arise in which a membrane stress produced by pressure or other mechanical loading and associated with a primary or a discontinuity effect produces excessive distortion in the transfer of load to other portions of the structure. Conservatism requires that such a stress be classified as local primary membrane stress even though it shows some characteristics of a secondary stress. Examples include the membrane stress in a shell produced by external loads and moment at a permanent support or at a nozzle connection.
   c. Bending stress, P'_b. This stress is the variable component of normal stress in a cross section. An example is the bending stress in the central portion of a flat head due to pressure.
2. Secondary stress, Q'. Secondary stress is a normal stress or a shear stress developed by the constraint of adjacent material or by self-constrain of the structure, and thus it is normally associated with deformation-controlled quantity. The basic characteristic of a secondary stress is that it is self-limiting. Local yielding and minor distortions can satisfy the conditions that cause the stress to occur, and failure from one application of the stress is not to be expected. Examples of secondary stresses are bending stress at a gross structural discontinuity, bending stress due to a linear radial thermal strain profile through the thickness of sections, and stress produced by an axial temperature distribution in a cylindrical shell.
3. Peak stress, F'. Peak stress is that increment of stress that is additive to the primary plus secondary stresses by reason of local geometry including stress concentrations. The basic characteristic of a peak stress is that it does not cause any noticeable distortion and is objectionable only as a possible source of a fatigue crack or a brittle fracture. Some examples of peak stress are thermal stress at a local discontinuities and cladding, thermal stress in the wall of a vessel caused by rapid change in temperature of the contained fluid, and stress at a structural discontinuity.

ASME Sections I and VIII give design rules for common components such as shells, heads, nozzles, and covers. These rules are intended to keep the stresses P'_m, P'_L, and P'_b in the components within allowable stress limits. The design pressure and temperature are assumed slightly higher than the operational pressure and temperature and are taken at a point in the operational cycle where they are a maximum. Many pressurized equipment such as power boilers, nuclear pressure components, and hydro crackers in refineries operate at essentially a steady-state condition. However, the combination of mechanical and thermal loading necessitates additional analysis. Section VIII-2 has a specific procedure for setting limits on various combinations of the stresses P'_m, P'_L, P'_b, and Q'. The maximum values of these combinations that are called equivalent stress values in Section VIII-2 (also called stress intensity values in Section III) are designated by P_m, P_L, P_b, and Q as shown in Figure 3.5. The procedure for calculating equivalent stress values from stress values is explained next.

The ASME code lists, in various tables and figures, extensive examples of stress categories in vessel components including their location, origin of stress, type of stress, and classification. The stress limits of primary, secondary, and peak stress categories are detailed in various sections of the codes such as the one shown in Figure 3.5, which is taken from Section VIII-2. The procedure for determining the quantities P_m, P_L, P_b, Q, and F in Figure 3.5 is as follows:

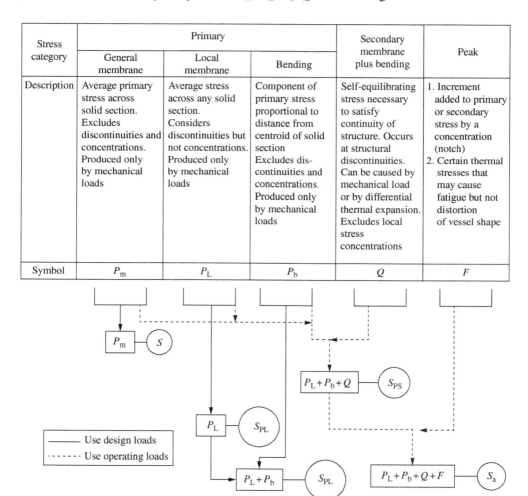

Stress category	Primary			Secondary membrane plus bending	Peak
	General membrane	Local membrane	Bending		
Description	Average primary stress across solid section. Excludes discontinuities and concentrations. Produced only by mechanical loads	Average stress across any solid section. Considers discontinuities but not concentrations. Produced only by mechanical loads	Component of primary stress proportional to distance from centroid of solid section Excludes discontinuities and concentrations. Produced only by mechanical loads	Self-equilibrating stress necessary to satisfy continuity of structure. Occurs at structural discontinuities. Can be caused by mechanical load or by differential thermal expansion. Excludes local stress concentrations	1. Increment added to primary or secondary stress by a concentration (notch) 2. Certain thermal stresses that may cause fatigue but not distortion of vessel shape
Symbol	P_m	P_L	P_b	Q	F

——— Use design loads
- - - - - Use operating loads

P_m — S

P_L — S_{PL}

$P_L + P_b + Q$ — S_{PS}

$P_L + P_b$ — S_{PL}

$P_L + P_b + Q + F$ — S_a

Figure 3.5 Stress categories and limits. Source: ASME, VIII-2.

1. At a given time during the operation (such as steady-state condition), separate the calculated stress at a given point in a vessel into (P'_m), $(P'_L + P'_b)$, and $(P'_L + P'_b + Q')$. The quantity P'_L is either general or local primary membrane stress.
2. Each of these three bracketed quantities is actually six quantities acting as normal and shear stresses on an infinitesimal cube at the point selected.
3. Repeat steps 1 through 2 at a different time frame such as start-up.
4. Find the algebraic sum of the stresses from steps 1 through 3. This sum is the stress range of the quantities (P'_m), $(P'_L + P'_b)$, and $(P'_L + P'_b + Q')$.
5. From step 4 determine the three principal stresses S_1, S_2, and S_3 for each of the stress categories (P'_m), $(P'_L + P'_b)$, and $(P'_L + P'_b + Q')$.
6. From step 5 determine the stress values of $S_1 - S_2$, $S_2 - S_3$, and $S_3 - S_1$ for each of the stress categories (P'_m), $(P'_L + P'_b)$, and $(P'_L + P'_b + Q')$.
7. Calculate the effective stress for each of the quantities calculated in step 6 that are now called equivalent stress (P_m), $(P_L + P_b)$, and $(P_L + P_b + Q)$. The effective stress is calculated as follows.

   The effective stress at a point and is given by the following:
   For Section VIII-2,

$$\sigma_e = 0.707 \left[(S_1 - S_2)^2 + (S_2 - S_3)^2 + (S_3 - S_1)^2 \right]^{0.5} \tag{3.24}$$

For Section III,

$$\text{Larger of the quantities } S_1 - S_2, S_2 - S_3, \text{ and } S_3 - S_1 \tag{3.25}$$

One limitation of the aforementioned two equations is they give essentially zero stress under hydrostatic load condition where $S_1 \approx S_2 \approx S_3$. Accordingly, the following equation is also applicable:

$$S_1 + S_2 + S_3 < 4\ S \tag{3.26}$$

where S is the allowable stress.

8. The equivalent stress values in step 7 are analyzed in accordance with Figure 3.5.

The design equivalent stress limits for P_m, P_L, and P_b are shown by the solid lines in Figure 3.5. The design equivalent stress is generally limited to

$$P_m < S_m \text{ for general membrane stress}$$
$$P_L < 1.5\ S_m \text{ for local membrane stress}$$
$$(P_L + P_b) < 1.5\ S_m \text{ for local membrane and bending primary stresses.}$$

The previously mentioned 1.5 shape factor is based on plastic theory (Jawad and Jetter 2011) and is applicable to rectangular cross sections such as local bending in a shell or plate. For bending of shells in a beam mode such as tubes in a heat exchanger, the shape factor can be as low as 1.14.

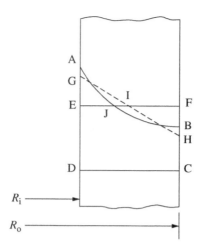

Figure 3.6 Decomposition of stresses in a cylindrical shell

For the operating condition, it is necessary to determine the equivalent stress values of P_m, P_L, P_b, and Q. The criteria in VIII-2 are given by the dotted lines in Figure 3.5 and show that primary (mainly pressure) plus secondary (mainly temperature) equivalent stress values are limited by the quantity S_{PS}. This quantity is essentially limited by the smaller of $3S$ or $2S_y$.

Many of the construction details given in VIII-2 for major components meet these design stress limits for internal pressure, and generally there is no need to run a stress check due to design conditions. Other components such as nozzle reinforcement, lugs, and attachments or thermal design conditions may require an additional analysis in accordance with Figure 3.5 when the design specifications require it or when the designer deems it necessary.

Based on the established categories described previously, the stress pattern in Figure 3.3 can be decomposed into membrane, bending, and peak stress as shown in Figure 3.6. These three stress values are

$$P'_m = \frac{p_i}{\gamma - 1} = \frac{PR_i}{t} \tag{3.27}$$

where
$\gamma = R_o/R_i$
p_i = internal pressure
t = thickness

$$Q' = \frac{6P\gamma}{(\gamma - 1)^2} \left[0.5 - \frac{\gamma}{\gamma^2 - 1} (\ln \gamma) \right] \tag{3.28}$$

$$F' = P \left(\frac{\gamma^2 + 1}{\gamma^2 - 1} \right) - P'_m - Q'. \tag{3.29}$$

Example 3.1

A pressure vessel with $r_i = 12$ inch is subjected to a pressure of 9615 psi. Determine

1. The required thickness using $S = 25,000$ psi
2. The values of p'_m, Q', and F'

Solution

1. From Eq. (3.22),

$$Z = \frac{25,000 + 9615}{25,000 - 9615} = 2.25$$

$$t = 12(2.25^{0.5} - 1) = 6.00 \text{ inch}$$

The maximum circumferential stress is at the inside surface of the shell.

$$\gamma = \frac{18}{12} = 1.5$$

$$\sigma_{\theta max} = 9615 \left(\frac{1.5^2 + 1}{1.5^2 - 1} \right) = 25,000 \text{ psi}$$

2. From Eq. (3.27),

$$P'_m = \frac{10,000(12)}{6.00} = 20,000 \text{ psi}$$

From Eq. (3.28),

$$Q' = 4650 \text{ psi}$$

From Eq. (3.29),

$$F' = 350 \text{ psi}$$

The elastic circumferential stress distribution through the thickness is shown in Figure 3.7.

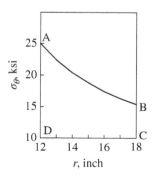

Figure 3.7 Circumferential stress distribution using elastic Eq. (3.20).

Example 3.2

A pressure vessel with an inside diameter of 72 inches, a tangent to tangent length of 18 ft, and a thickness of 1.75 inches is subjected to a design pressure of 970 psi and operating pressure of 875 psi. The vessel is supported by four legs exerting a shear stress in the shell at the vicinity of the legs of 7000 psi due to vessel weight. The axial stress in the shell due to vessel weight is −220 psi. Determine the maximum stress in the shell and compare it with the allowable stress value of 20,000 psi. Assume the shell to be thin and let the yield stress equal to 38,000 psi.

Solution

Condition 1 when pressure is zero:

    Circumferential membrane stress = 0 psi

    Longitudinal membrane stress = −220 psi

    Radial membrane stress = 0 psi

    Shear stress = 7000 psi

Condition 2 when design pressure is 970 psi:

    Circumferential membrane stress = PR/t = (970)(36)/1.75 = 19,950 psi

    Longitudinal membrane stress = $(PR/2t)$ − 220 = 9975 − 220 = 9755 psi

    Radial membrane stress = − 970/2 = − 485 psi

    Radial bending stress = ±485 psi

    Shear stress = 7000 psi

Condition 3 when operating pressure is 875 psi:

    Circumferential membrane stress = PR/t = (875)(36)/1.75 = 18,000 psi

    Longitudinal membrane stress = $(PR/2t)$ − 220 = 9000 − 220 = 8780 psi

    Radial membrane stress = −875/2 = − 438 psi

    Radial bending stress = ±438 psi

    Shear stress = 7000 psi

Check primary stresses:

    From Figure 3.5, primary stresses are obtained from the design condition only:

P_{mc} = 19 950 − 0 = 19,950 psi
P_{ml} = 9755 − (−220) = 9975 psi
P_{mr} = −485 − 0 = −485 psi
P_{cl} = 7000 − 7000 = 0 psi

These are principal stresses since the shearing stress P_{cl} is zero.

Hence, the membrane stress P_m used in Figure 3.5 is obtained from Eq. (3.24) as

$$P_m = 0.707 \left[(19,950 - 9975)^2 + (9975 - (-485))^2 + (-485 - 19,950)^2 \right]^{1/2}$$

P_m = 17,700 psi

From Figure 3.5, this value is less than the allowable stress of 20,000 psi.

Check primary plus secondary stresses at the inside surface:

    From Figure 3.5, primary plus secondary stresses are obtained from the operating condition only:

$$P_{mc} + Q_c = (18,000 - 0) + (0 - 0) = 18,000 \, psi$$
$$P_{ml} + Q_l = [8780 - (-220)] + (0 - 0) = 9000 \, psi$$
$$P_{mr} + Q_r = (-438 - 0) + (-438 - 0) = -875 \, psi$$
$$P_{cl} = 7000 - 7000 = 0 \, psi$$

These are principal stresses since the shearing stress P_{cl} is zero.

Hence, the primary plus secondary stress $(P_m + Q)$ used in Figure 3.4 is obtained from Eq. (3.24) as

$$P_m + Q = 0.707 \left[(18{,}000 - 9000)^2 + [9000 - (-875)]^2 + (-875 - 18{,}000)^2 \right]^{1/2}$$

$$P_m + Q = 16{,}350 \, \text{psi}$$

This value is less than the allowable stress permitted in Figure 3.5 (lesser of $3S$ or $2S_y$).

3.4 Plastic Analysis of Thick Wall Cylinders

Figure 3.3 shows that the maximum circumferential stress due to internal pressure is at the inside surface of the cylinder. As the internal pressure increases, the stress at the inside surface reaches the yield stress. The internal pressure needed to cause yield stress can be determined from calculating the shear stress at the inner surface. Combining Eqs. (3.12) and (3.13) and setting $r = R_i$ gives

$$\sigma_\theta - \sigma_r = 2p_i R_o^2 / (R_o^2 - R_i^2). \tag{3.30}$$

Tresca's criterion for the yield envelope is expressed as

$$\tau_{max} = \frac{\sigma_y}{2} = \frac{\sigma_\theta - \sigma_r}{2}. \tag{3.31}$$

Combining Eqs. (3.29) and (3.30) and solving for p_i result in

$$p_i = (\sigma_y/2)(R_o^2 - R_i^2)/R_o^2. \tag{3.32}$$

Equation (3.32) determines the pressure at which yield first occurs at the inner surface of the cylinder. Further increase in internal pressure causes the yield stress to propagate through the thickness as shown in Figure 3.8.

The pressure required to yield the total thickness can be determined from Eq. (3.1):

$$\sigma_\theta - \sigma_r = r d\sigma_r / dr. \tag{3.33}$$

The left-hand side of this equation, divided by two, gives the shearing stress. Hence, combining Eqs. (3.31) and (3.33) gives

$$d\sigma_r / dr = (1/r)(\sigma_y). \tag{3.34}$$

The solution of this equation is

$$\sigma_r = \sigma_y \ln r + C_1. \tag{3.35}$$

The value of C_1 is obtained from the boundary condition at the outside surface of the cylinder where $\sigma_r = 0$ at $r = R_o$. Thus,

$$C_1 = -\sigma_y \ln R_o. \tag{3.36}$$

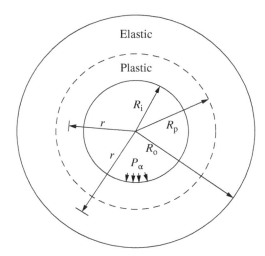

Figure 3.8　Elastic–plastic regions in a cylinder due to internal pressure.

The radial stress is obtained from Eqs. (3.35) and (3.36) as

$$\sigma_r = \sigma_y \ln\left(\frac{r}{R_o}\right). \tag{3.37}$$

The stress at the inner surface is

$$\sigma_r = \sigma_y \ln\left(\frac{R_i}{R_o}\right)$$

and the internal pressure required to bring the total thickness into a plastic flow condition is

$$p_i = -\sigma_y \ln\left(\frac{R_i}{R_o}\right). \tag{3.38}$$

Equation (3.38) determines the pressure at which total yielding occurs in the wall of the cylinder. This equation is often used for design purposes by using allowable stress S in lieu of σ_y, design internal pressure P in lieu of p_i, and $R_o = R_i + t$. This gives

$$t = R_i\left(e^{P/S} - 1\right). \tag{3.39}$$

The circumferential stress in the cylinder is obtained by combining Eqs. (3.31) and (3.37):

$$\sigma_\theta = \sigma_y\left[1 + \ln\left(\frac{r}{R_o}\right)\right].$$

For design purposes, this equation is often written as

$$S_\theta = S\left[1 + \ln\left(\frac{r}{R_o}\right)\right] \tag{3.40}$$

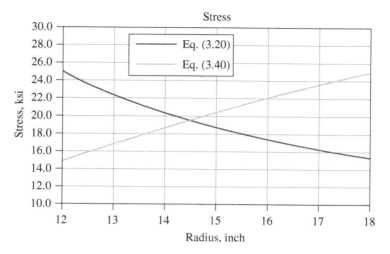

Figure 3.9 Circumferential stress distribution using elastic Eq. (3.20) and Plastic Eq. (3.40)

where
S_θ = actual circumferential stress
S = allowable stress

Example 3.3
What is the required thickness of the shell in Example 3.1 using plastic theory?

Solution
From Eq. (3.39),

$$t = 12 \, (e^{9615/25,000} - 1)$$
$$t = 5.62 \text{ inch}$$

This thickness is about 7% less than that obtained in Example 3.1 from elastic theory. Figure 3.9 shows stress distribution of σ_θ through the thickness.

3.5 Creep Analysis of Thick-Wall Cylinders

The expression for stress due to internal pressure in a thick cylinder operating in the creep range is based on Norton's equation:

$$\dot{\varepsilon}* = K'\sigma*^{n} \tag{3.41}$$

where
$\dot{\varepsilon}*$ = effective strain rate
E = modulus of elasticity
K' = constant (when $n = 1$, $K' = 1/E$)
n = creep exponent, which is a function of material property and temperature
$\sigma*$ = effective stress

Equation (3.1) for summation of forces in the r direction is valid in the creep range and is expressed as

$$\sigma_\theta - \sigma_r - r\, d\sigma_r / dr = 0. \tag{3.42}$$

The effective stress, $\sigma*$, is based on Tresca's shear stress criterion with a value obtained from Eq. (3.31) as

$$\frac{\sigma*}{2} = \frac{\sigma_\theta - \sigma_r}{2}. \tag{3.43}$$

Substituting this equation and Eq. (3.33), for the equilibrium of forces in a cross section, into Norton's Eq. (3.41) gives

$$\dot{\varepsilon}* = K'(r\, d\sigma_r / dr)^n. \tag{3.44}$$

The expression for the effective strain is developed next. The strain–deflection relationships in a cylinder are obtained from strength of materials:

$$\varepsilon_\theta = \frac{u}{r} \quad \text{or} \quad \dot{\varepsilon}_\theta = \frac{\dot{u}}{r} \tag{3.45}$$

$$\varepsilon_r = \frac{du}{dr} \quad \text{or} \quad \dot{\varepsilon}_r = \frac{d\dot{u}}{dr}. \tag{3.46}$$

It is also assumed that the volume of the material does not change in the creep range. Hence,

$$\varepsilon_r + \varepsilon_\theta + \varepsilon_L = 0. \tag{3.47}$$

For a long cylinder, the value of ε_L does not change with length. Hence, ε_L is constant and is normally taken to be zero. Equation (3.46) becomes

$$\varepsilon_r = -\varepsilon_\theta \quad \text{or} \quad \dot{\varepsilon}_r = -\dot{\varepsilon}_\theta. \tag{3.48}$$

Substituting Eqs. (3.45) and (3.46) into Eq. (3.48) gives

$$\frac{d\dot{u}}{dr} = \frac{-\dot{u}}{r}.$$

A solution of this equation is

$$\dot{u} = \frac{-C_1}{r}.$$

Hence, the values of ε_θ and ε_r become

$$\varepsilon_\theta = \frac{-C_1}{r^2} \quad \varepsilon_r = \frac{C_1}{r^2}.$$

Tresca's yield envelope is applicable to strain limits as

$$\frac{\varepsilon*}{2} = \frac{\varepsilon_\theta - \varepsilon_r}{2} \quad \text{or} \quad \frac{\dot{\varepsilon}*}{2} = \frac{\dot{\varepsilon}_\theta - \dot{\varepsilon}_r}{2}. \tag{3.49}$$

Hence,

$$\dot{\varepsilon}* = \frac{-2C_1}{r^2}.$$

Substituting this expression into Eq. (3.44) and rearranging terms results in

$$d\sigma_r = \frac{1}{r^{(2/n)+1}}(-2C_1/K')^{1/n}dr$$

$$d\sigma_r = C_2\left(r^{-(2/n)-1}\right)dr.$$

(3.50)

Integrating this expression gives

$$\sigma_r = C_2(-2/n)^{-1}\left(r^{-(2/n)}\right) + C_3.$$

(3.51)

Constants C_2 and C_3 are obtained from the boundary conditions $\sigma_r = 0$ at $r = R_o$ and $\sigma_r = -P_i$ at $r = R_i$:

$$C_2 = \frac{-2P_i}{n\left(R_o^{-2/n} - R_i^{-2/n}\right)}$$

$$C_3 = \frac{-P_i R_o^{-2/n}}{\left(R_o^{-2/n} - R_i^{-2/n}\right)}.$$

Substituting C_2 and C_3 back into Eq. (3.51) gives the value of σ_r as

$$\sigma_r = -p_i \frac{(R_o/r)^{2/n} - 1}{(R_o/R_i)^{2/n} - 1}.$$

(3.52)

The value of σ_θ is obtained from Eq. (3.41) as

$$\sigma_\theta = p_i \frac{[(2-n)/n](R_o/r)^{2/n} + 1}{(R_o/R_i)^{2/n} - 1}.$$

(3.53)

The longitudinal stress σ_L is obtained from the stress–strain relationship:

$$\varepsilon_L = \frac{1}{E}(\sigma_L - \mu\sigma_r - \mu\sigma_\theta).$$

(3.54)

For long cylinders, the value of ε_L can be taken as zero as discussed previously. Also, the value of μ in the creep range is assumed to be 0.5. Hence, the longitudinal stress is

$$\sigma_L = 0.5(\sigma_r + \sigma_\theta)$$

and is the average of the σ_r and σ_θ stresses:

$$\sigma_L = p_i \frac{[(1-n)/n](R_o/r)^{2/n} + 1}{(R_o/R_i)^{2/n} - 1}$$

(3.55)

The aforementioned equations are normally applied to stationary problems where the operation is mainly steady state rather than transient. For this condition it can be shown (Jawad and Jetter 2011) that Norton's Eq. (3.41) may be replaced (Hult 1966) with a modified equation of the form

$$\varepsilon = K'\sigma^n. \tag{3.56}$$

This equation is used extensively in creep analysis and is also used to determine the values of n needed in Eqs. (3.52), (3.53), an3.55.55). The value of n is usually obtained from isochronous curves for a specific material at a given temperature for a desired time period. As an illustration, the isochronous curves for 2.25Cr-1Mo annealed steel are shown in Figure 3.10 at 1000°F for

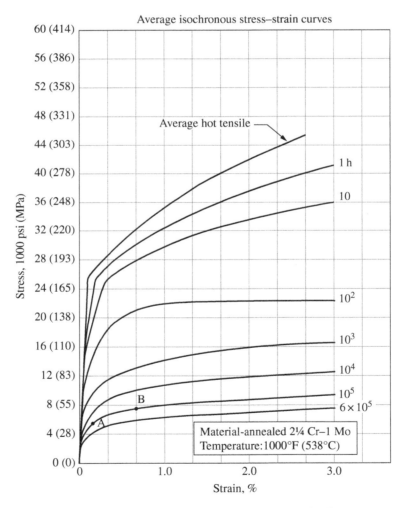

Figure 3.10   Isochronous stress–strain curves (ASME)

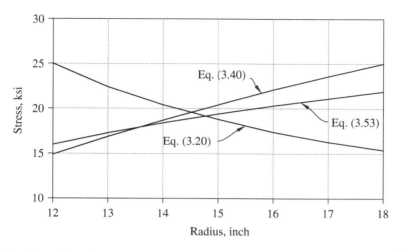

Figure 3.11 Circumferential stress distribution using elastic Eq. (3.20), plastic Eq. (3.40), and creep Eq. (3.53)

various time periods. For the 100,000 hour curve, two points, A and B, are arbitrarily chosen as shown in Figure 3.10. The values of stress and strain corresponding to these two points are obtained from the figure. These values are substituted into Eq. (3.56) to determine the values of n and K', resulting in $n = 5.1$ and $K' = 1.049 \times 10^{-22}$.

Example 3.4
Find the maximum stress of the shell in Example 3.1 using creep theory. Let $n = 5.1$.

Solution
From Eq. (3.53), it is seen that the maximum stress occurs at $r = R_o$. Hence,

$$\sigma_\theta = (9615)\frac{[(2-5.1)/5.1](18/18)^{2/5.1} + 1}{(18/12)^{2/5.1} - 1}$$

$$\sigma_\theta = 21{,}900 \, \text{psi}.$$

The stress distribution through the thickness is shown in Figure 3.11.

3.6 Shell Equations in the ASME Code

3.6.1 Boiler Code: Section I

Section I lists Lame's Eq. (3.22) in paragraph A-125. It also lists the following empirical equation in paragraph PG-27.2.2:

$$t = \frac{PR}{SE - (1-y)P} \tag{3.57}$$

where

E = weld joint efficiency
P = internal pressure
R = inside radius
S = allowable stress
t = thickness
y = creep factor which is a function of temperature and material

Equation (3.57) approximates Lame's Eq. (3.22) as well as the creep Eq. (3.53). Thus, when $y = 0.4$, the equation reduces to

$$t = \frac{PR}{SE-0.6P}.$$ (3.58)

This equation gives essentially the same results as Lame's Eq. (3.22). This can be verified by comparing the results obtained from Example 3.1 with that obtained from this equation (with $E = 1$).

When $y = 0.7$, Eq. (3.57) becomes

$$t = \frac{PR}{SE-0.3P}.$$ (3.59)

This equation gives essentially the same results as the creep Eq. (3.53). This can be verified by comparing the results obtained from Example 3.4 with that obtained from this equation (with $E = 1$).

3.6.2   Nuclear Code: Section III

Paragraph NB-3324.1 lists the following equation for establishing the minimum thickness in the shell:

$$t = \frac{PR}{S_m-0.5P}$$ (3.60)

where S_m = allowable stress.

This equation is obtained from the principal stress theory, which states that the maximum stress is equal to the largest difference in the three principal stresses: longitudinal, circumferential, and radial. The membrane stresses in the three principal exes for a thin shell are

$$S_c = \frac{PR}{t} \quad S_L = \frac{PR}{2t}, \quad \text{and} \quad S_r = \frac{-P}{2}.$$

And the maximum principal stress is $S_m = PR/t - (-P/2)$. This expression can be rearranged to give Eq. (3.60).

3.6.2.1   Subsections NC, ND, and NE

Lame's equation is listed in paragraphs NC-3224.3, ND-3224.3, and NE-3224.3 which also list the following simplification of Lame's equation:

$$t = \frac{PR}{S-0.6P}$$ (3.61)

where S is allowable stress.

3.6.3 Pressure Vessel Code: Section VIII

3.6.3.1 Division 1 (ASME 2017a)

The simplified Lame's Eq. (3.58) is listed in paragraph UG-27 of VIII-1. Also, paragraphs 1–2 in Appendix 1 of VIII-1 list the plastic Eq. (3.39) as an alternate design equation for thick cylinders.

3.6.3.2 Division 2 (ASME 2017b)

Paragraph 4.3.3.1 uses the plastic Eq. (3.39) as a basis for design.

3.6.4 Basis of Simplified Lame's Equations in the ASME Code

The constant 0.6 appears in the denominator of Eq. (3.58) for the boiler code and Eq. (3.61) for the nuclear code, Divisions NC, ND, and NE, and in the pressure vessel code, Division 1. It also changes to a value of 0.3 in Eq. (3.59) for boilers at elevated temperatures. The rationale for this variation is based on the assumed application of these equations as follows:

Let

$$t = \frac{PR}{S - XP}$$

where X is an unknown coefficient to be determined. This equation can be written as

$$S = P(R + Xt)/t. \tag{3.62}$$

Equating this equation to Lame's Eq. (3.21a) and solving for X give

$$X = \frac{(R/t) + 1}{(2R/t) + 1}. \tag{3.63}$$

For heavy wall cylinders where the thickness approaches one-half the radius, $R/t = 2$ and $X = 0.6$. Hence, the denominator of Eqs. (3.58) and (3.61) is indexed to a value of 0.6 since R/t can be as high as 2.0 in such vessels.

For temperatures in the creep range, assuming $n = 5.1$, the value of X is obtained by solving Eqs. (3.62) and (3.53):

$$X = \frac{0.385}{(1 + t/R)^{0.385} - 1} - R/t.$$

For heavy wall cylinders where $R/t \approx 2.0$, Eq. (3.63) gives $X = 0.28$.

From Eq. (3.57), $(y - 1) = X$ or $y = 0.72$, which is practically the same as the value listed in the boiler code, Section I for carbons steels at elevated temperatures.

3.7 Bending of Thin-Wall Cylinders Due to Axisymmetric Loads

3.7.1 Basic Equations

The membrane forces discussed in the last two chapters are sufficient to resist many commonly encountered loading conditions. At locations where the deflection is restricted or there is a change in geometry such as cylindrical-to-spherical shell junction, the membrane theory is

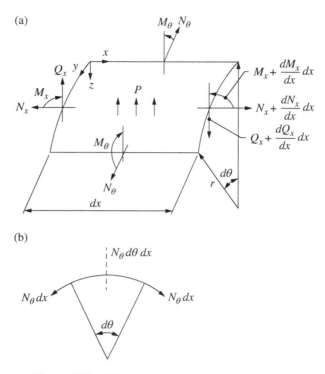

Figure 3.12   Infinitesimal section of a cylinder shell

inadequate to maintain deflection and rotation compatibility between the shells as illustrated in
Example 2.3. At these locations discontinuity forces are developed, which result in bending and
shear stresses in the shell. These discontinuity forces are localized over a small area of the shell
and dissipate rapidly along the shell. Many structures such as pressure vessels and storage tanks
are designed per the membrane theory, and the total stress at discontinuities is determined from
the membrane and bending theories. In this chapter the bending theory of cylindrical shells is
developed, and in Chapter 5 the bending theory of spherical and conical shells is discussed.
Other references for the analysis of cylindrical and spherical shells include Kraus (1967),
Timoshenko and Woinowsky-Krieger (1959), and Ugural (1998).

We begin the derivation of the bending of thin cylindrical shells by assuming the applied
loads to be symmetric with respect to angle θ. A free-body diagram of an infinitesimal
section of a cylindrical shell is shown in Figure 3.12. The radius of the cylinder is designated
as r. The applied loads p can vary in the x-direction only. At edges $x = 0$ and $x = dx$, the axial
membrane force N_x, bending moments M_x, and shearing forces Q_x are axisymmetric. In the
circumferential direction, only the hoop membrane force N_θ and bending moments M_θ are
needed for equilibrium. There are no shearing forces, Q_θ, because the applied loads are symmetric in the circumferential direction. Summation of forces in the x-direction gives the first
equation of equilibrium:

$$(N_x r d\theta) - \left(N_x + \frac{dN_x}{dx}dx\right)r d\theta = 0$$

or

$$\frac{dN_x}{dx}r\,dx\,d\theta = 0. \tag{3.64}$$

This equation indicates that N_x must be a constant. We assume a cylinder with open ends and set $N_x = 0$. In Section 3.7.3, we will discuss the case where N_x is not zero.

Summation of forces in the z-direction gives the second equation of equilibrium:

$$Q_x r\,d\theta - \left(Q_x + \frac{dQ_x}{dx}dx\right)r\,d\theta - N_\theta\,dx\,d\theta + pr\,dx\,d\theta = 0$$

or

$$\frac{dQ_x}{dx} + \frac{N_\theta}{r} = p. \tag{3.65}$$

Summation of moments around the y-axis gives the third equation of equilibrium:

$$M_x r\,d\theta - \left(M_x + \frac{dM_x}{dx}dx\right)r\,d\theta + \left(Q_x + \frac{dQ_x}{dx}dx\right)(r\,d\theta)\,dx$$

$$-pr\,d\theta\frac{dx}{2}dx + 2N_\theta\frac{d\theta}{2}dx\frac{dx}{2} = 0.$$

After simplifying and deleting terms of higher order, we get

$$\frac{dM_x}{dx} - Q_x = 0. \tag{3.66}$$

Eliminating Q_x from Eqs. (3.65) and (3.66) gives

$$\frac{d^2 M_x}{dx^2} + \frac{N_\theta}{r} = p. \tag{3.67}$$

Equation (3.67) contains two unknowns, N_θ and M_x. Both of these unknowns can be expressed in terms of deflection, w. Define axial strain as

$$\varepsilon_x = \frac{du}{dx}. \tag{3.68}$$

The circumferential strain is obtained from Figure 3.13 as

$$\varepsilon_\theta = \frac{2\pi(r+\Delta r)-2\pi r}{2\pi r}$$

$$\varepsilon_\theta = \frac{\Delta r}{r}$$

or

$$\varepsilon_\theta = \frac{-w}{r} \tag{3.69}$$

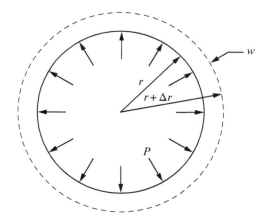

Figure 3.13 Circumferential deflection

where w is the deflection and is taken as positive inwards. The stress–strain relationship is obtained from the theory of elasticity and can be written in terms of force–strain relationship as

$$\begin{bmatrix} N_x \\ N_\theta \end{bmatrix} = \frac{Et}{1-\mu^2} \begin{bmatrix} 1 & \mu \\ \mu & 1 \end{bmatrix} \begin{bmatrix} \varepsilon_x \\ \varepsilon_\theta \end{bmatrix}. \tag{3.70}$$

Notice that the shearing strain, $\gamma_{x\theta}$, is zero in this case due to load symmetry in the θ-direction. Substituting Eqs. (3.68) and (3.69) into the first expression of Eq. (3.70) results in

$$N_x = \frac{Et}{1-\mu^2} \left(\frac{du}{dx} - \mu \frac{w}{r} \right).$$

Substituting into this expression the value $N_x = 0$ from Eq. (3.64), we get

$$\frac{du}{dx} = \mu \frac{w}{r}. \tag{3.71}$$

Similarly, the second term of Eq. (3.70) can be written as

$$N_\theta = \frac{Et}{1-\mu^2} \left(-\frac{w}{r} + \mu \frac{du}{dx} \right),$$

or upon inserting Eq. (3.71), it becomes

$$N_\theta = -\frac{Etw}{r}. \tag{3.72}$$

The basic moment–deflection relationships of Eq. (7.17) for flat plates are also applicable to thin cylindrical shells. Referring to the two axes as x and θ rather than x and y, the first two expressions in Eq. (7.17) become

$$\begin{bmatrix} M_x \\ M_\theta \end{bmatrix} = -D \begin{bmatrix} 1 & \mu \\ \mu & 1 \end{bmatrix} \begin{bmatrix} \dfrac{d^2 w}{dx^2} \\ \dfrac{d^2 w}{d\theta^2} \end{bmatrix}.$$ (3.73)

It should be noted that x and θ in Eq. (3.73) are not in polar coordinates but redefined x- and y-axes. Polar transformation of Eq. (3.73) is given in Chapter 4.

The third expression, M_{xy}, in Eq. (7.17) vanishes because the rate of change of deflection with respect to θ is zero due to symmetry of applied loads. Also, due to symmetry with respect to θ, all derivatives with respect to θ vanish, and the first expression in Eq. (3.73) reduces to

$$M_x = -D \frac{d^2 w}{dx^2}$$ (3.74)

and the second expression in Eq. (3.73) becomes

$$M_\theta = -\mu D \frac{d^2 w}{dx^2}.$$ (3.75)

From Eqs. (3.74) and (3.75), it can be concluded that

$$M_\theta = \mu M_x.$$ (3.76)

Substituting Eqs. (3.72) and (3.74) into Eq. (3.67) gives

$$\frac{d^4 w}{dx^4} + \frac{Et}{Dr^2} w = -p(x)/D$$

which is the differential equation for the bending of cylindrical shells due to loads that are variable in the x-direction and uniformly distributed in the θ-direction.

Defining

$$\beta^4 = \frac{Et}{4Dr^2} = \frac{3(1-\mu^2)}{r^2 t^2}$$ (3.77)

the differential equation becomes

$$\frac{d^4 w}{dx^4} + 4\beta^4 w = -p(x)/D$$ (3.78)

where p is a function of x.

Solution of Eq. (3.78) results in an expression for the deflection, w. The longitudinal and circumferential moments are then obtained from Eqs. (3.74) and (3.76), respectively. The circumferential membrane force, N_θ, is determined from Eq. (3.72).

One solution of Eq. (3.78) that is commonly used for long cylindrical shells is expressed as

$$w = e^{\beta x}(C_1 \cos \beta x + C_2 \sin \beta x) + e^{-\beta x}(C_3 \cos \beta x + C_4 \sin \beta x) + f(x)$$ (3.79)

where $f(x)$ is the particular solution and C_1–C_4 are constants that are evaluated from the boundary conditions.

A different solution of Eq. (3.78) that is commonly used for short cylindrical shells is expressed as

$$w = C_5 \sin \beta x \sinh \beta x + C_6 \sin \beta x \cosh \beta x,$$
$$+ C_7 \cos \beta x \sinh \beta x + C_8 \cos \beta x \cosh \beta x. \tag{3.80}$$

The procedure for establishing moments and forces in cylindrical shells as well as defining long and short cylinders is discussed in the following sections.

For sign convention, M_x, Q_x, N_θ, and M_θ are all positive as shown in Figure 3.12. The deflection, w, is positive inward and the rotation, ψ, in the x-direction is positive in the direction of positive bending moments.

3.7.2 Long Cylindrical Shells

One application of Eq. (3.79) is of shear forces and bending moments applied at the edge of a cylindrical shell, Figure 3.14. Referring to Eq. (3.79) for the deflection of a shell, we can set the function $f(x)$ to zero as there are no applied loads along the cylinder. Also, the deflection due to the term $e^{\beta x}$ in Eq. (3.79) tends to approach infinity as x gets larger. However, the deflection due to moments and forces applied at one end of an infinitely long cylinder tend to dissipate as x gets larger. Thus, constants C_1 and C_2 must be set to zero, and Eq. (3.79) becomes

$$w = e^{-\beta x}(C_3 \cos \beta x + C_4 \sin \beta x). \tag{3.81}$$

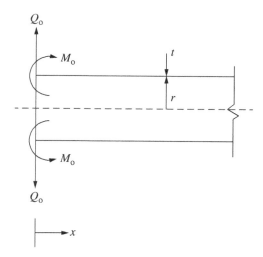

Figure 3.14 Shear and moment applied at the edge of cylinder

The boundary conditions for the infinitely long cylinder, Figure 3.14, are obtained from Eq. (3.74) as

$$M_x|_{x=0} = M_0 = -D\frac{d^2w}{dx^2}\bigg|_{x=0}$$

and from Eq. (3.66) as

$$Q_x|_{x=0} = Q_0 = \frac{dM}{dx}\bigg|_{x=0}.$$

Substituting Eq. (3.81) into the first boundary condition gives

$$C_4 = \frac{M_0}{2\beta^2 D}$$

and from the second boundary condition

$$C_3 = -\frac{1}{2\beta^3 D}(Q_0 + \beta M_0).$$

Hence, the deflection equation for the long cylinder shown in Figure 3.14 is

$$w = \frac{e^{-\beta x}}{2\beta^3 D}[M_0(\sin\beta x - \cos\beta x) - Q_0\cos\beta x]. \tag{3.82}$$

By defining

$$A_{\beta x} = e^{-\beta x}(\cos\beta x + \sin\beta x) \tag{3.83}$$

$$B_{\beta x} = e^{-\beta x}(\cos\beta x - \sin\beta x) \tag{3.84}$$

$$C_{\beta x} = e^{-\beta x}\cos\beta x \tag{3.85}$$

$$D_{\beta x} = e^{-\beta x}\sin\beta x \tag{3.86}$$

the expression for the deflection and its derivative becomes

$$\text{Deflection} = w_x = \frac{-1}{2\beta^3 D}\left(\beta M_0 B_{\beta x} + Q_0 C_{\beta x}\right) \tag{3.87}$$

$$\text{Slope} = \psi_x = \frac{1}{2\beta^2 D}\left(2\beta M_0 C_{\beta x} + Q_0 A_{\beta x}\right) \tag{3.88}$$

$$\text{Moment} = M_x = \frac{1}{2\beta}\left(2\beta M_0 A_{\beta x} + 2Q_0 D_{\beta x}\right) \tag{3.89}$$

$$\text{Shear} = Q_x = -\left(2\beta M_0 D_{\beta x} - Q_0 B_{\beta x}\right). \tag{3.90}$$

The functions $A_{\beta x}$ through $D_{\beta x}$ are calculated in Table 3.1 for various values of βx.

Table 3.1   Values of functions $A_{\beta x}$, $B_{\beta x}$, $C_{\beta x}$, and $D_{\beta x}$

βx	$A_{\beta x}$	$B_{\beta x}$	$C_{\beta x}$	$D_{\beta x}$
0	1.0000	1.0000	1.0000	0.0000
0.05	0.9976	0.9025	0.9500	0.0475
0.10	0.9907	0.8100	0.9003	0.0903
0.15	0.9797	0.7224	0.8510	0.1286
0.20	0.9651	0.6398	0.8024	0.1627
0.30	0.9267	0.4888	0.7077	0.2189
0.40	0.8784	0.3564	0.6174	0.2610
0.50	0.8231	0.2415	0.5323	0.2908
0.60	0.7628	0.1431	0.4530	0.3099
0.80	0.6354	−0.0093	0.3131	0.3223
1.00	0.5083	−0.1108	0.1988	0.3096
1.20	0.3899	−0.1716	0.1091	0.2807
1.40	0.2849	−0.2011	0.0419	0.2430
1.60	0.1959	−0.2077	−0.0059	0.2018
1.80	0.1234	−0.1985	−0.0376	0.1610
2.00	0.0667	−0.1794	−0.0563	0.1231
2.50	−0.0166	−0.1149	−0.0658	0.0491
3.00	−0.0423	−0.0563	−0.0493	0.0070
3.5	−0.0389	−0.0177	−0.0283	−0.0106
4.0	−0.0258	0.0019	−0.0120	−0.0139
5.0	−0.0045	0.0084	0.0019	−0.0065
6.0	0.0017	0.0031	0.0024	−0.0007
7.0	0.0013	0.0001	0.0007	0.0006

Example 3.5

A long cylindrical shell is subjected to end moment M_o as shown in Figure 3.15a. Plot the value of M_x from $\beta x = 0$ to $\beta x = 5.0$. Also determine the distance x at which the moment is about 6.7% of the original applied moment M_o.

Solution

From Eq. (3.89),

$$M_x = M_o A_{\beta x}.$$

The values of $A_{\beta x}$ are obtained from Eq. (3.83) and a plot of M_x is shown in Figure 3.15b. From Eq. (3.89) and Table 3.1, the value of βx at which M_x is equal to about 6.7% of M_o is about 2.0. Hence,

$$\beta x = 2.0$$

or

$$x = 1.56\sqrt{rt} \quad \text{for} \quad \mu = 0.3.$$

The significance of the quantity $1.56 \sqrt{rt}$ is apparent from Figure 3.15b. It shows that a moment applied at the end of a long cylinder dissipates very rapidly as x increases and it reduces to about 6.7% of the original value at a distance of $1.56 \sqrt{rt}$ from the edge.

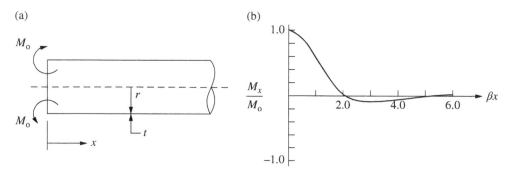

Figure 3.15 Long cylinder with edge moment M_o: (a) edge moment and (b) moment distribution along the length

Many design codes use similar procedure to that shown in Example 3.5 to establish effective length of cylinders. The following are some effective lengths listed in Section VIII-1 of the ASME Code.

Paragraph	Equation	Effective length of
UG-29	$1.56\sqrt{rt}$	Stiffener–shell composite section
UHX-12.5.10	$2.55\sqrt{rt}$	Shell welded to tubesheet
1-5(e)	$2.0\sqrt{rt}$	Shell welded to large end of cone
1-5(e)	$1.4\sqrt{rt}$	Shell welded to small end of cone

Example 3.6
Determine the maximum stress in a long cylinder due to the radial load shown in Figure 3.16a.

Solution
From the free-body diagram of Figure 3.16b, the end load at point A is equal to $Q_o/2$. Also, from symmetry the slope is zero at point A, Thus,

$$\text{Slope due to } M_o + \text{slope due to } \frac{Q_o}{2} = 0.$$

From Eq. (3.88),

$$\frac{M_o}{\beta D} - \frac{Q_o}{4\beta^2 D} = 0$$

or

$$M_o = \frac{Q_o}{4\beta}.$$

The maximum longitudinal bending

$$\text{Stress} = 6M_o/t^2 = \frac{3Q_o}{2\beta t^2}$$

$$\text{Maximum deflection at point } A = \frac{-M_o}{2\beta^2 D} + \frac{Q_o}{4\beta^3 D} = \frac{Q_o}{8\beta^3 D}.$$

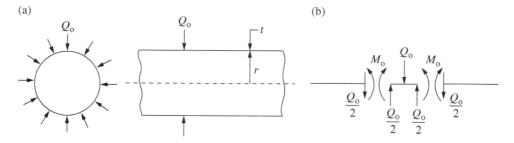

Figure 3.16   Long cylinder with a concentrated radial load Q_o: (a) radial load on a cylinder and (b) free-body diagram

The circumferential membrane force is obtained from Eq. (3.72) using a positive sign for tension:

$$N_\theta = \frac{Etw}{r} = \frac{EtQ_o}{8r\beta^3 D}.$$

The circumferential bending moment is

$$M_\theta = \mu M_x$$

and the total maximum circumferential stress using a positive sign for tension is

$$\sigma_\theta = \frac{EQ_o}{8r\beta^3 D} + \frac{3\mu Q_o}{2\beta t^2}.$$

Example 3.7

Determine the expression for stress in the long cylinder shown in Figure 3.17a due to an internal pressure of 100 psi. The cylinder is supported by rigid bulkheads. Assume the longitudinal force N_x to be zero.

Solution

For a long cylindrical shell, Eq. (3.79) can be written as

$$w = e^{-\beta x}(C_3 \cos \beta x + C_4 \sin \beta x) + f(x).$$

The particular solution for a constant pressure along the length of the cylinder can be expressed as

$$w_p = K.$$

Substituting this expression in Eq. (3.78) results in

$$w_p = -\frac{p}{4\beta^4 D}$$

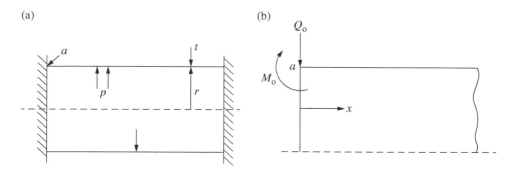

Figure 3.17 Long cylinder with fixed edges: (a) cylindrical shell fixed at ends and (b) discontinuity forces

which can also be expressed as

$$w_p = -\frac{pr^2}{Et}.$$

A free-body diagram of the discontinuity forces at the end is shown in Figure 3.17b. The unknown moment and shear forces are assumed in a given direction as shown. A negative answer will indicate that the true direction is opposite the assumed one. The compatibility condition requires that the deflection at the edge due to pressure plus moment plus shear is equal to zero. Referring to Eq. (3.87) and Figure 3.17,

$$\frac{-p}{4\beta^4 D} - \frac{M_o}{2\beta^2 D} + \frac{Q_o}{2\beta^3 D} = 0$$

or

$$M_o - \frac{Q_o}{\beta} = \frac{-p}{2\beta^2}. \qquad (1)$$

Similarly the slope due to pressure plus moment plus shear is equal to zero at the edge. Hence, from Eq. (3.87) and Figure 3.17,

$$0 + \frac{M_o}{\beta D} - \frac{Q_o}{2\beta^2 D} = 0$$

or

$$M_o - \frac{Q_o}{2\beta} = 0. \qquad (2)$$

Solving Eqs. (1) and (2) gives

$$M_o = \frac{p}{2\beta^2} \quad \text{and} \quad Q_o = \frac{p}{\beta}.$$

Thus, the deflection is given by Eq. (3.87) and Figure 3.17 as

$$w = \frac{p}{2\beta^4 D}\left(-\frac{1}{2}B_{\beta x} + C_{\beta x}\right) - \frac{p}{4\beta^4 D}.$$

At the bulkhead attachment, the circumferential membrane force N_θ is zero because the deflection is zero in accordance with Eq. (3.72). The axial bending moment is given by Eq. (3.74) as

$$M_x = -D\frac{d^2w}{dx^2} = \frac{pe^{-\beta x}}{2\beta^2}(\sin\beta x - \cos\beta x).$$

The circumferential bending moment is given by Eq. (3.76) as

$$M_\theta = \mu M_x = \frac{\mu p e^{-\beta x}}{2\beta^2}(\sin\beta x - \cos\beta x).$$

The circumferential membrane force along the cylinder is given by Eq. (3.72) as

$$N_\theta = \frac{-Et}{r}\left(\frac{p}{2\beta^4 D}\right)(-B_{\beta x} + 2C_{\beta x} - 1).$$

Longitudinal bending stress $\sigma_{Lb} = \pm 6M_x/t^2$
Longitudinal membrane stress $\sigma_{Lm} = 0$
Circumferential bending stress $\sigma_{\theta b} = \mu\sigma_{Lb}$
Circumferential membrane stress $\sigma_{\theta m} = N_\theta/t$

It is of interest to note that if a shearing force Q (Figure 3.14) is applied at the end of a cylinder by itself, then the bending moment is zero at that location and the maximum bending moment is away from the point of application of the shearing force. This occurs in situations such as shown in Figure 3.18 where a cylinder is filled with fluid up to a certain point such that the temperature below the fluid level is different than that above the fluid level. In such cases the bending moment is zero at the interface and only shear forces are present. Thus, the maximum bending moment occurs away from the location of discontinuity.

Care must also be exercised in cases where the discontinuity forces are applied close together as shown in Figures 3.19 and 3.20. In such cases the discontinuity stresses at one end are influenced by forces from both ends.

3.7.3 Long Cylindrical Shells with End Loads

The radial deflection obtained from Eq. (3.81) is based on the assumption that the axial membrane force, N_x, in Eq. (3.64) is negligible. However, many applications involve pressure and hydrostatic loads that result in axial forces. The deflections and slopes due to these axial forces must be determined first, and then the deflections and slopes due to edge effects described in the previous section are superimposed for a final solution. This procedure is illustrated here for a cylindrical shell with end closures and subjected to internal pressure.

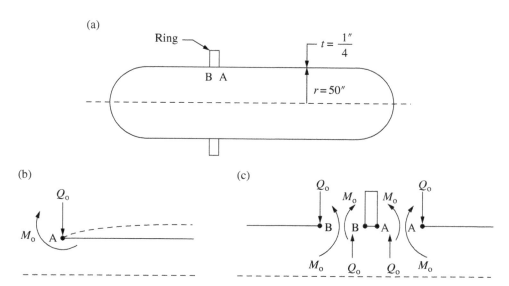

Figure 3.18 Pressure vessel with stiffening ring: (a) cylindrical shell, (b) discontinuity forces, and (c) free-body diagram

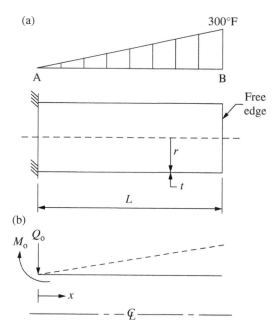

Figure 3.19 Temperature distribution along a cylinder: (a) cylindrical shell with axial temperature distribution and (b) discontinuity forces

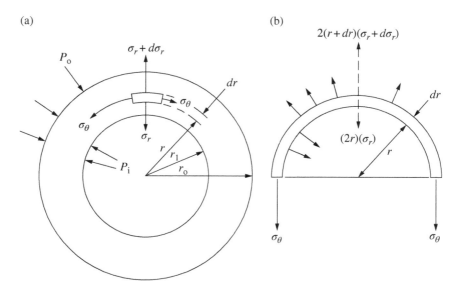

Figure 3.20 Circumferential and radial stress in a cylinder: (a) thick cylinder and (b) infinitesimal section

Let a cylindrical shell be subjected to internal pressure p. Then the circumferential, hoop, stress at a point away from the ends is obtained from Eq. (1.30) as

$$\sigma_\theta = \frac{pr}{t}. \tag{3.91}$$

Similarly the longitudinal stress is obtained from Eq. (1.32) as

$$\sigma_x = \frac{pr}{2t}. \tag{3.92}$$

The maximum stresses given by Eqs. (3.91) and (3.92) for thin cylindrical shells subjected to internal pressure are valid as long as the shell is allowed to grow freely. Any restraints such as end closures and stiffening rings that prevent the shell from growing freely will result in bending moments and shear forces in the vicinity of the restraints. The magnitude of these moments and forces is determined subsequent to solving Eq. (3.81). The total stress will then be a summation of those obtained from Eqs. (3.91) and (3.92) plus those determined as a result of solving Eq. (3.81).

The circumferential and axial strains are obtained from Eqs. (3.2) (3.3), (3.91), and (3.92) as

$$\varepsilon_\theta = \frac{pr}{Et}(1-\mu/2) \tag{3.93}$$

$$\varepsilon_x = \frac{pr}{2Et}(1-2\mu). \tag{3.94}$$

The radial deflection is obtained from Figure 3.13 as

$$\varepsilon_\theta = \frac{2\pi(r+\Delta r)-2\pi r}{2\pi r}$$

$$= \frac{\Delta r}{r} = \frac{-w}{r}$$

or

$$w = -r\varepsilon_\theta.$$

From this expression and Eq. (3.93), we get

$$w = -\frac{pr^2}{Et}(1-\mu/2). \tag{3.95}$$

Example 3.8

A stiffening ring is placed around a cylinder at a distance removed from the ends as shown in Figure 3.18a. The radius of the cylinder is 50 inches and its thickness is 0.25 inch. Also, the internal pressure = 100 psi, $E = 30,000$ ksi, and $\mu = 0.3$. Find the discontinuity stresses if (a) the ring is assumed to be infinitely rigid and (b) the ring is assumed to be 4 inches wide × 3/8 inch thick.

Solution

a.

$$\beta = \sqrt[4]{\frac{3(1-0.3^2)}{50^2 \times 0.25^2}} = 0.36357 \; 1/\text{inch}.$$

$$D = \frac{30,000,000 \times 0.25^3}{12(1-0.3^2)} = 42,925.82 \; \text{lbs-inch}.$$

From Eq. (3.95),

$$w_p = \frac{100 \times 50^2}{30,000,000 \times 0.25}(1-0.3/2) = 0.0283 \, \text{inch}.$$

From Figure 3.18b, it is seen that for an infinitely rigid ring, the deflection is zero. Also from symmetry, the slope is zero, and Eqs. (3.87) and (3.88) give

$$w_M = \frac{M_o}{2\beta^2 D} \quad \theta_M = \frac{M_o}{\beta D}$$

$$w_Q = \frac{Q_o}{2\beta^3 D} \quad \theta_Q = \frac{Q_o}{2\beta^2 D}.$$

Total deflection at the ring attachment is equal to zero:

$$0.0283 + \frac{M_o}{2\beta^2 D} - \frac{Q_o}{2\beta^3 D} = 0. \tag{1}$$

Similarly, the slope at the ring attachment is zero:

$$\frac{M_o}{\beta D} - \frac{Q_o}{2\beta^2 D} = 0. \tag{2}$$

From Eq. (2),

$$Q_o = 2\beta M_o.$$

From Eq. (1),

$$M_o = 321.53 \, \text{inch-lbs/inch}$$

and

$$Q_o = 233.79 \, \text{lbs/inch}.$$

Stress in the ring is zero because it is infinitely rigid:
Longitudinal bending stress in cylinder $= 6M_o/t^2 = 30{,}870$ psi
Circumferential bending stress $= 0.3 \times 30{,}870 = 9260$ psi
Circumferential membrane stress at ring junction $= 0$ psi

$$\text{Total longitudinal stress} = 30{,}870 + \frac{100 \times 50}{2 \times 0.25} = 40{,}870 \, \text{psi}$$

Total circumferential stress $= 9260$ psi

b. From Figure 3.18c, and symmetry, we can conclude that the shear and moment in the shell to the left of the ring are the same as the shear and moment to the right of the ring. Accordingly, we can solve only one unknown shear and one unknown moment value by taking the discontinuity forces of a shell on one side of the ring only. The deflection of the ring due to pressure can be ignored because the ring's width is 16 times that of the shell. The deflection of the ring due to $2Q_o$ is given by

$$w_R = \frac{2Q_o r^2}{AE}.$$

Compatibility of the shell and ring deflections require that

$$\text{Deflection of shell} = \text{deflection of ring}$$

or, from Figure 3.18c,

$$(\text{Deflection due to } p - \text{deflection due to } Q + \text{deflection due to } M)|_{\text{shell}}$$
$$= \text{deflection due to } 2Q_o|_{\text{ring}}$$

$$\frac{pr^2}{Et}(1 - \mu/2) - \frac{Q_o}{2\beta^3 D} + \frac{M_o}{2\beta^2 D} = \frac{2Q_o r^2}{AE} \tag{3}$$

$$M_o - 4.01 Q_o = -321.4.$$

From symmetry, the rotation of the shell due to pressure plus Q_o plus M_o must be set to zero:

$$\frac{Q_o}{2\beta^2 D} = \frac{M_o}{\beta D}$$

or

$$Q_o = 2\beta M_o. \tag{4}$$

Solving Eqs. (3) and (4) gives

$$Q_o = 122\,\text{lbs/inch}$$
$$M_o = 167.7\,\text{inch-lbs/inch}.$$

Notice that this moment is about half of the moment in the case of an infinitely rigid ring:

$$\text{Stress in ring} = \frac{2Q_o r}{A} = 8130\,\text{psi}.$$

Maximum longitudinal stress in shell occurs at the ring attachment and is given by

$$\sigma_x = \frac{pr}{2t} + \frac{6M}{t^2} = 10,000 + 16,100 = 26,100\,\text{psi}.$$

Deflection of shell at ring junction is given by

$$w = w_p - w_Q + w_M$$

$$w = \frac{pr^2}{Et}(0.85) - \frac{Q_o}{2\beta^3 D} + \frac{M_o}{2\beta^2 D}$$

$$w = \frac{406,120}{E}\,\text{inch}.$$

The circumferential membrane force is

$$N_\theta = \frac{Etw}{(1-\mu^2)r} = 2231.4\,\text{ibs/inch}.$$

The circumferential bending moment is

$$M_\theta = \mu M_x = 0.3 \times 167.7 = 50.3\,\text{inch-lbs/inch}$$

$$\sigma_\theta = \frac{2231.4}{0.25} + \frac{6 \times 50.3}{0.25^2} = 8930 + 4830 = 13,760\,\text{psi}.$$

3.7.4   Short Cylindrical Shells

It was shown in the previous sections that the deflection due to applied edge-shearing forces and bending moments dissipates rapidly as x increases, and it becomes negligible at distances larger than $2.0/\beta$. This rapid reduction in deflection as x increased simplifies the solution of Eq. (3.79) by letting $C_1 = C_2 = 0$. When the length of the cylinder is less than about $2.0/\beta$, then C_1 and C_2 cannot be ignored and all four constants in Eq. (3.79) must be evaluated. Usually the alternate equation for the deflection, Eq. (3.80), results in a more convenient solution for short cylinders than Eq. (3.79). The calculations required in solving Eq. (3.80) are tedious because four constants are evaluated rather than two.

Example 3.9

Derive N_θ due to applied bending moment M_o at edge $x = 0$ for a short cylinder of length L.

Solution

The four boundary conditions are

At $x = 0$

$$\text{Moment} = M_\text{o} = -D\left(\frac{d^2 w}{dx^2}\right)$$

$$\text{Shear} = 0 = -D\left(\frac{d^3 w}{dx^3}\right).$$

At $x = L$

$$\text{Moment} = 0 = -D\left(\frac{d^2 w}{dx^2}\right)$$

$$\text{Shear} = 0 = -D\left(\frac{d^3 w}{dx^3}\right).$$

The second derivative of Eq. (3.80) is

$$\frac{d^2 w}{dx^2} = 2\beta^2 (C_1 \cos \beta x \cosh \beta x + C_2 \cos \beta x \sinh \beta x \tag{1}$$

$$- C_3 \sin \beta x \cosh \beta x - C_4 \sin \beta x \sinh \beta x).$$

The third derivative of Eq. (3.80) is

$$\frac{d^3 w}{dx^3} = 2\beta^3 [C_1 (\cos \beta x \sinh \beta x - \sin \beta x \cosh \beta x)$$

$$+ C_2 (\cos \beta x \cosh \beta x - \sin \beta x \sinh \beta x)$$

$$- C_3 (\sin \beta x \sinh \beta x + \cos \beta x \cosh \beta x)$$

$$- C_4 (\sin \beta x \cosh \beta x + \cos \beta x \sinh \beta x)]. \tag{2}$$

Substituting Eq. (1) into the first boundary condition gives

$$C_1 = \frac{-M_\text{o}}{2D\beta^2}.$$

Substituting Eq. (2) into the second boundary condition gives

$$C_2 = C_3.$$

From the third and fourth boundary conditions, we obtain

$$C_3 = \frac{M_o}{2D\beta^2}\left(\frac{\sin\beta L\cos\beta L + \sinh\beta L\cosh\beta L}{\sinh\beta L - \sin\beta L}\right)$$

and

$$C_4 = \frac{M_o}{2D\beta^2}\left(\frac{\sin\beta L + \sinh\beta L}{\sinh\beta L - \sin\beta L}\right).$$

From Eq. (3.72),

$$N_\theta = \frac{Etw}{r}$$

$$= \frac{Et}{r}(C_1\sin\beta x\sinh\beta x + C_2\sin\beta x\cosh\beta x$$

$$+ C_3\cos\beta x\sinh\beta x + C_4\cos\beta x\cosh\beta x).$$

3.8   Thermal Stress

3.8.1   *Stress Due to Thermal Gradient in the Axial Direction*

Thermal temperature gradients in cylindrical shells occur either along the axial length or through the thickness of cylinders. Thermal gradients through the thickness are discussed in Section 3.8.2. Stress in a cylindrical shell due to temperature gradients in the axial direction can be obtained by subdividing the cylinder into infinitesimal rings of length dx. The thermal expansion in each ring due to a change of temperature T_x within the ring is given by

$$w = \alpha r T_x$$

where α is the coefficient of thermal expansion. Some values of α are shown in Table 3.2.

Since adjacent cylindrical rings cannot have a mismatch in the deflection due to temperature T_x at their interface, an assumed pressure p_x must be applied to eliminate the temperature deflection mismatch. Hence,

$$\frac{p_x r^2}{Et} = \alpha r T_x$$

and

$$p_x = Et\alpha T_x/r$$

$$\sigma_\theta = -p_x r/t = -E\alpha T_x. \tag{3.96}$$

As the cylinder does not have any actual applied loads on it, the forces p_x must be eliminated by applying equal and opposite forces to the cylinder. Hence, Eq. (3.78) becomes

$$\frac{d^4 w}{dx^4} + 4\beta^4 w = \frac{Et\alpha T_x}{rD}. \tag{3.97}$$

The total stress in a cylinder due to axial thermal gradient distribution is obtained by adding the stresses obtained from Eqs. (3.96) and (3.97).

Table 3.2 Coefficients of thermal expansion (multiplied by 10^6), inch/inch/°F

Material	Room temperature	200	400	600	800	1000
Aluminum	12.6	12.9	13.5			
Brass (Cu–Zn)	9.6	9.7	10.2	10.7	11.2	11.6
Bronze (Cu–Al)				9.0^a		
Carbon steel	6.5	6.7	7.1	7.4	7.8	
Copper	9.4	9.6	9.8	10.1	10.3	10.5
Cu–Ni (70–30)	8.5	8.5	8.9	9.1		
Nickel alloy C276	6.1	6.3	6.7	7.1	7.3	
Nickel alloy 600	6.9	7.2	7.6	7.8	8.0	
Stainless steel	8.6	8.8	9.2	9.5	9.8	10.1
Titanium (Gr. 1,2)	4.7	4.7	4.8	4.9	5.1	
Zirconium alloys	3.2	3.4	3.7	4.0		

Temperature, °F

^{a} At 500°F.

Example 3.10

The cylinder shown in Figure 3.19a is initially at 0°F. The cylinder is heated as shown in the figure. Determine the thermal stresses in the cylinder. Let $\alpha = 6.5 \times 10^{-6}$ inch/inch/°F, $E = 30,000$ ksi, $L = 10$ ft, $t = 0.25$ inch, $r = 30$ inches, and $\mu = 0.3$. The cylinder is fixed at point A and free at point B.

Solution

The temperature gradient is expressed as $T_x = 300\ x/L$. The circumferential stress due to ring action is given by Eq. (3.96) as

$$\sigma_\theta = -E\alpha 300 x/L. \tag{1}$$

Equation (3.97) is written as

$$\frac{d^4 w}{dx^4} + 4\beta^4 w = \frac{E t \alpha}{rD}(300 x/L).$$

A particular solution of this equation in taken as

$$w = C_1 x + C_2.$$

Substituting this expression into the differential equation gives

$$C_1 = 300 r\alpha/L \quad C_2 = 0$$

and

$$w = -r\alpha(300 x/L).$$

From Eq. (3.72),

$$\sigma_\theta = \frac{N_\theta}{t} = \frac{Ew}{r}. \tag{2}$$

Adding Eqs. (1) and (2) gives

$$\sigma_\theta = 0$$

which indicates that the thermal stress in a cylinder due to linear axial thermal gradient is zero. The slope, Figure 3.19b, due to thermal gradient is

$$\theta = \frac{dw}{dx} = r\alpha \left(\frac{300}{L}\right).$$

At the fixed end, bending moments will occur due to the rotation θ caused by thermal gradients. The boundary conditions at the fixed edge are $w = 0$ and $\theta = 0$. From Eq. (3.87) and Figure 3.19b,

$$0 = \beta M_\mathrm{o} + Q_\mathrm{o}.$$

From Eq. (3.88),

$$r\alpha \left(\frac{300}{L}\right) = \frac{1}{2\beta^2 D}(2\beta M_o + Q_o).$$

Solving these two equations gives

$$M_\mathrm{o} = 2r\alpha(300)\beta D/L$$

$$= 19.63 \,\text{inch-lbs/inch}$$

$$\sigma = 6M_\mathrm{o}/t^2 = 6 \times 19.63/0.25^2 = 1880 \,\text{psi}.$$

3.8.2 Stress Due to Thermal Gradient in the Radial Direction

The governing equations for the thermal stress in a cylinder due to temperature gradient in the radial direction are obtained by combining various equations derived previously. It is assumed that all shearing strains are zero and that the strain due to temperature is given by αT where α is the coefficient of thermal expansion and T is the temperature. It is also assumed that the axial strain ε_x is uniform and constant. The elastic stress–strain equations can be written as (Faupel and Fisher 1981)

$$\epsilon_\mathrm{r} = \frac{1}{E}[\sigma_\mathrm{r} - \mu(\sigma_\theta + \sigma_z)] + \alpha T \tag{3.98}$$

$$\epsilon_\theta = \frac{1}{E}[\sigma_\theta - \mu(\sigma_z + \sigma_\mathrm{r})] + \alpha T \tag{3.99}$$

$$\epsilon_z = \frac{1}{E}[\sigma_z - \mu(\sigma_\mathrm{r} + \sigma_\theta)] + \alpha T. \tag{3.100}$$

The circumferential strain is obtained from Eq. (3.69) as

$$\varepsilon_\theta = \frac{w}{r} \tag{3.101}$$

where outward radial deflection, w, is taken as positive. Radial strain is obtained from Eq. (3.4) as

$$\varepsilon_r = \frac{dw}{dr}. \tag{3.102}$$

Equations (3.101) and (3.102) are combined to give

$$\varepsilon_r = \varepsilon_\theta + r\left(\frac{d\varepsilon_\theta}{dr}\right). \tag{3.103}$$

Substituting Eqs. (3.98) and (3.99) into Eq. (3.103) yields

$$\sigma_r = \sigma_\theta + r\left(\frac{d\sigma_r}{dr}\right) - \frac{\mu}{1-\mu}\left(\sigma_r + r\frac{d\sigma_r}{dr} - \sigma_\theta\right) + \left(\frac{E}{1-\mu^2}\right)\alpha(1+\mu)r\left(\frac{dT}{dr}\right). \tag{3.104}$$

The relationship between circumferential and radial stress in a thick cylinder is shown in Figure 3.21 from Eq. (3.1):

$$\sigma_\theta - \sigma_r = r\left(\frac{d\sigma_r}{dr}\right). \tag{3.105}$$

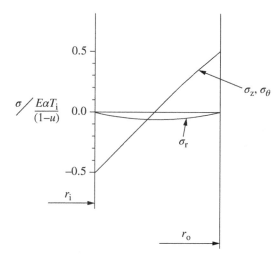

Figure 3.21   Thermal stress distribution in a wall of a cylinder

Combining Eqs. (3.104) and (3.105) and solving for σ_r result in

$$r\frac{d}{dr}\left[\frac{1}{r}\frac{d}{dr}(r^2\sigma_r)\right] = \frac{-E}{1-\mu^2}\alpha(1+\mu)r\left(\frac{dT}{dr}\right). \tag{3.106}$$

Solving this differential equation and applying the boundary conditions

$$\sigma_r|_{r=r_i} = 0$$
$$\sigma_r|_{r=r_o} = 0$$

give

$$\sigma_r = \frac{E\alpha}{1-\mu^2}\frac{1+\mu}{r^2}\left(\frac{r^2-r_i^2}{r_o^2-r_i^2}\int_{r_i}^{r_o} Tr\,dr - \int_{r_i}^{r} Tr\,dr\right). \tag{3.107}$$

From Eq. (3.105),

$$\sigma_\theta = \frac{E\alpha}{1-\mu^2}\frac{1+\mu}{r^2}\left(\frac{r^2+r_i^2}{r_o^2-r_i^2}\int_{r_i}^{r_o} Tr\,dr + \int_{r_i}^{r} Tr\,dr - Tr^2\right) \tag{3.108}$$

and from Eq. (3.100) for a cylinder unrestrained in the z-direction ($\epsilon_z = \alpha T$),

$$\sigma_z = \frac{E\alpha}{1-\mu}\left(\frac{2}{r_o^2-r_i^2}\int_{r_i}^{r_o} Tr\,dr - T\right). \tag{3.109}$$

From these three equations, some cases can be derived.

Case 1   Thin Shells with Linear Thermal Distribution

For thin vessels, a steady-state temperature condition produces linear thermal distribution through the thickness that can be expressed as

$$T = T_i\frac{r_o-r}{r_o-r_i} \tag{3.110}$$

where $T_i =$ inside wall temperature relative to outside wall temperature. Substituting T into Eqs. (3.107) through (3.109) gives

$$\sigma_r = \frac{E\alpha T_i}{r^2(1-\mu)}\left[\frac{(r^2-r_i^2)(2r_i+r_o)}{6(r_i+r_o)} - \frac{2(r^3-r_i^3)-3r_o(r^2-r_i^2)}{6(r_i-r_o)}\right] \tag{3.111}$$

$$\sigma_\theta = \frac{E\alpha T_i}{r^2(1-\mu)}\left[\frac{(r^2+r_i^2)(2r_i+r_o)}{6(r_i+r_o)} - \frac{2(r^3+r_i^3)-3r_o(r^2+r_i^2)}{6(r_i-r_o)}\right] \tag{3.112}$$

$$\sigma_z = \frac{E\alpha T_i}{(1-\mu)}\left[\frac{2r_i + r_o}{3(r_i + r_o)} - \frac{r_o - r}{r_o - r_i}\right]. \tag{3.113}$$

Figure 3.21 is a typical plot of σ_r, σ_θ, and σ_z. The plot indicates that σ_r is small compared with σ_θ and σ_z, and for all practical purposes, σ_θ and σ_z are equal.

The maximum values of σ_θ and σ_z occur at the inner and outer surfaces.

From Eqs. (3.111) and (3.112),

$$\sigma_\theta = \sigma_z = \begin{cases} \dfrac{-E\alpha T_i}{1-\mu}\left[\dfrac{2r_o + r_i}{3(r_o + r_i)}\right] & \text{for inside surface} \\[3mm] \dfrac{E\alpha T_i}{1-\mu}\left[\dfrac{r_o + 2r_i}{3(r_o + r_i)}\right] & \text{for outside surface} \end{cases}. \tag{3.114}$$

For thin-wall vessels, Eq. (3.114) reduces to

$$\sigma_\theta = \sigma_z = \begin{cases} \dfrac{-E\alpha T_i}{2(1-\mu)} & \text{for inside surface} \\[3mm] \dfrac{E\alpha T_i}{2(1-\mu)} & \text{for outside surface} \end{cases}. \tag{3.115}$$

Case 2 Thick Shells with Logarithmic Thermal Distribution

In thick vessels, a steady-state thermal condition gives rise to a logarithmic temperature distribution (Burgreen 1971) that can be expressed as

$$T = T_i\left(\frac{\ln r_o - \ln r}{\ln r_o - \ln r_i}\right). \tag{3.116}$$

Substitution of this expression in Eqs. (3.107) through 3.1099) results in

$$\sigma_r = \frac{-E\alpha T_i}{2(1-\mu)\ln(r_o/r_i)}\left[\ln\frac{r_o}{r} + \frac{r_i^2}{r_o^2 - r_i^2}\left(1 - \frac{r_o^2}{r^2}\right)\ln\left(\frac{r_o}{r_i}\right)\right] \tag{3.117}$$

$$\sigma_\theta = \frac{E\alpha T_i}{2(1-\mu)\ln(r_o/r_i)}\left[1 - \ln\frac{r_o}{r} - \frac{r_i^2}{r_o^2 - r_i^2}\left(1 + \frac{r_o^2}{r^2}\right)\ln\left(\frac{r_o}{r_i}\right)\right] \tag{3.118}$$

$$\sigma_z = \frac{E\alpha T_i}{2(1-\mu)\ln(r_o/r_i)}\left[1 - 2\ln\frac{r_o}{r} - \frac{2r_i^2}{r_o^2 - r_i^2}\ln\left(\frac{r_o}{r_i}\right)\right]. \tag{3.119}$$

Disregarding σ_r as being small compared to σ_θ and σ_z, Eqs. (3.117) and (3.119) have a maximum value of

$$\sigma_\theta = \sigma_z = \begin{cases} \dfrac{E\alpha T_i}{2(1-\mu)\ln(r_o/r_i)}\left[1 - \dfrac{2\,r_o^2}{r_o^2 - r_i^2}\ln\left(\dfrac{r_o}{r_i}\right)\right] & \text{for inside surface} \\[2em] \dfrac{E\alpha T_i}{2(1-\mu)\ln(r_o/r_i)}\left[1 - \dfrac{2\,r_i^2}{r_o^2 - r_i^2}\ln\left(\dfrac{r_o}{r_i}\right)\right] & \text{for outside surface} \end{cases} \tag{3.120}$$

And for thin-wall cylinders, Eq. (3.120) reduces to

$$\sigma_\theta = \sigma_z = \begin{cases} \dfrac{-E\alpha T_i}{2(1-\mu)} & \text{for inside surface} \\[2em] \dfrac{E\alpha T_i}{2(1-\mu)} & \text{for outside surface} \end{cases} \tag{3.121}$$

which are the same as those for the linear case.

Case 3   Thick Shells with Complex Thermal Distribution

In many applications, such as transient and upset conditions, the temperature distribution through the wall of a vessel cannot be represented by a mathematical expression. In this case a graphical solution can be obtained for the thermal stress. From Eq. (3.108),

$$\sigma_\theta = \frac{E\alpha}{1-\mu}\left[\frac{1+(r_i/r)^2}{r_o^2 - r_i^2}\int_{r_i}^{r_o} Tr\,dr + \frac{1}{r^2}\int_{r_i}^{r} Tr\,dr - T\right]. \tag{3.122}$$

For a cylinder where the thickness is small compared with the radius, the first expression in the brackets of Eq. (3.122) can be expressed as

$$\frac{1+(r_i/r)^2}{r_o^2 - r_i^2}\int_{r_i}^{r_o} Tr\,dr = \frac{2\pi\displaystyle\int_{r_i}^{r_o} Tr\,dr}{\pi\left(r_o^2 - r_i^2\right)}$$

$$= \text{mean value of the temperature}$$

$$\text{distribution through the wall thickness}$$

$$= T_m. \tag{3.123}$$

The second expression can be expressed as

$$\frac{1}{r^2} \int_{r_i}^r Tr\,dr = \frac{2\pi \int_o^r Tr\,dr}{2\pi r^2} \tag{3.124}$$

= one-half the mean value of the temperature
distribution from the axis of the vessel to r.

However, because the temperature distribution from the axis to r_i is zero, this latter expression for all practical purposes can be neglected. Hence, σ_θ can be expressed as

$$\sigma_\theta = \frac{E\alpha}{1-\mu}(T_m - T) \tag{3.125}$$

where, T_m = mean value of temperature distribution through the wall and T = temperature at desired location.

From Eq. (3.109), it can be seen that σ_z can also be approximated by Eq. (3.125).

Example 3.11
A thin cylindrical vessel is heated from the outside such that the temperature distribution is as shown in Figure 3.22. If $E = 29 \times 10^6$ psi, $\alpha = 9.3 \times 10^{-6}$ inch/inch-°F, and $\mu = 0.28$, determine the maximum thermal stress (a) using Eq. (3.114) and (b) using Eq. (3.115).

Solution
a. $T_i = 450 - 750 = -300°$F. Hence at the inside surface

$$\sigma = \frac{-\left(29 \times 10^6\right)\left(9.3 \times 10^{-6}\right)(-300)}{(1-0.28)}\left(\frac{2(13)+11}{3(13+11)}\right)$$

$$= 57,750\,\text{psi}$$

and at the outside surface

$$\sigma = \frac{\left(29 \times 10^6\right)\left(9.3 \times 10^{-6}\right)(-300)}{(1-0.28)}\left(\frac{13 + 2 \times 11}{3(13+11)}\right)$$

$$= 54,630\,\text{psi}.$$

b. At the inside surface

$$\sigma = \frac{\left(-29 \times 10^6\right)\left(9.3 \times 10^{-6}\right)(-300)}{2(1-0.28)}$$

$$= 56,200\,\text{psi}$$

and at the outside surface $\sigma = -56,200$ psi.

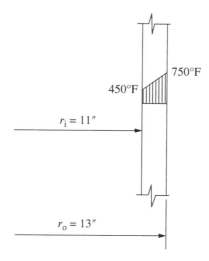

Figure 3.22   Temperature distribution in a thin-wall vessel

Example 3.12
A pressure vessel operating at 400°F is subjected to a short excursion temperature of 700°F. At a given time, the temperature distribution in the wall is shown in Figure 3.23. Find the maximum thermal stress at that instance. Let $\mu = 0.3$, $E = 30 \times 10^6$ psi, and $\alpha = 6.0 \times 10^{-6}$ inch/inch °F.

Solution
The maximum stress is determined from Eq. (3.125). The mean temperature is obtained from Figure 3.23 and Table 3.3.
   From Eq. (3.125) at the inner surface,

$$\sigma = \frac{\left(30 \times 10^6\right)\left(6.0 \times 10^{-6}\right)}{1 - 0.3}(455.5 - 700)$$

$$= -62,900 \, \text{psi}$$

and at the outer surface,

$$\sigma = \frac{\left(30 \times 10^6\right)\left(6.0 \times 10^{-6}\right)}{1 - 0.3}(455.5 - 400)$$

$$= 14,300 \, \text{psi}.$$

   It is of interest to note that the high stress occurs at the surface only. Thus at one-10th of the thickness inside the surface, the stress is

$$\sigma = \frac{\left(30 \times 10^6\right)\left(6 \times 10^{-6}\right)}{1 - 0.3}(455.5 - 560)$$

$$= -26,900 \, \text{psi}.$$

The high stress at the inner surface indicates that local yielding will occur.

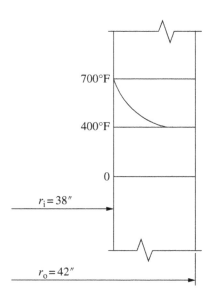

Figure 3.23   Temperature distribution in a thick-wall vessel

Table 3.3   Mean temperature

Location across, inch	Temperature, °F	Area[a], inch-°F
0.0	700	
0.4	560	252
0.8	500	212
1.2	470	194
1.6	440	182
2.0	420	172
2.4	410	166
2.8	405	163
3.2	400	161
3.6	400	160
4.0	400	160
Sum		1822
Mean T	455.5	

[a] Areas are approximated as rectangles and triangles.

3.9   Discontinuity Stresses

The design of various components in a shell structure subjected to axisymmetric loads consists of calculating the thickness of the main components first using the membrane theory and a given allowable stress. The forces due to various boundary conditions, such as those listed in Table 3.4, are then determined in accordance with the methods discussed in this chapter.

Table 3.4 Various discontinuity functions

	Edge functions	
w	$\dfrac{-M_o}{2\beta^2 D}$	$\dfrac{-H_o}{2\beta^3 D}$
θ	$\dfrac{-M_o}{\beta D}$	$\dfrac{H_o}{2\beta^2 D}$
M_x	M_o	0
N_θ	$2M_o\beta^2 r$	$2H_o\beta r$
Q_o	0	H_o
	General functions	
w	$\dfrac{-M_o}{2\beta^2 D}B_{\beta x}$	$\dfrac{-H_o}{2\beta^3 D}C_{\beta x}$
θ	$\dfrac{M_o}{\beta D}C_{\beta x}$	$\dfrac{H_o}{2\beta^2 D}A_{\beta x}$
M_x	$M_o A_{\beta x}$	$\dfrac{H_o}{\beta}D_{\beta x}$
N_θ	$2M_o\beta^2 r B_{\beta x}$	$2H_o\beta r C_{\beta x}$
Q_o	$2\beta M_o D_{\beta x}$	$H_o B_{\beta x}$

In most cases, the magnitude of the discontinuity bending and membrane stresses at the junction is high. However, these high stresses are very local in nature and dissipate rapidly away from the junction as shown in the examples previously solved. Tests and experience have shown that these stresses are secondary in nature and are allowed to exceed the yield stress without affecting the structural integrity of the components.

Many design codes such as the ASME pressure vessel and nuclear codes generally limit the secondary stresses at a junction to less than twice the yield stress at temperatures below the creep-rupture range. This stress level corresponds to approximately three times the allowable stress because the allowable stress is set at two-thirds of the yield stress value. The justification for limiting the stress to twice the yield stress is best explained by referring to Figure 3.24. The material stress–strain diagram is approximated by points ABO in Figure 3.24. In the first loading cycle, the discontinuity stress, calculated elastically, at the junction increases from point A to B and then to C as the applied load is increased. The secondary stress is allowed to approach twice the yield stress indicated by point C. This point corresponds to point D on the actual stress–strain diagram that is in the plastic range.

When the applied loads are reduced, the local discontinuity stress at point D is also reduced along the elastic line DEF. The high discontinuity stress at the junction is very localized in nature, and the material around the localized area is still elastic. Thus, when the applied loads

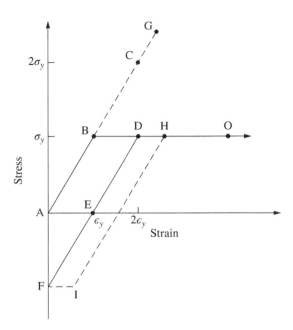

Figure 3.24   Stress–strain diagram

are reduced, the elastic material in the vicinity of the plastic region tends to return to its original zero strain and causes the much smaller volume of plastic material with high discontinuity stress to move from points D to E and then to F. Accordingly, at the end of the first cycle after the structure is loaded and then unloaded, the highly stressed discontinuity area that was stressed to twice the yield stress in tension is now stressed in compression with the yield stress value.

On subsequent loading cycles, the discontinuity stress is permitted to have a magnitude of twice the yield stress. However, the high stressed area that is now at point F moves to point E and then to point D. The high stress with a magnitude of twice the yield stress in the junction remains within the elastic limit on all subsequent loading cycles.

If the secondary discontinuity stress at the junction is allowed to exceed twice the yield stress such as point G, then for the first loading cycle, the strain approaches point H on the actual stress–strain diagram. Downloading will follow the path from H to I and then to F. Hence, yielding of the junction will occur both in the up and down cycles. Subsequent cycles will continue the yielding process that results in incremental plastic deformation at the junction that could lead to premature fatigue failure.

Problems

3.1 What is the required thickness of a cylindrical shell with $R = 15$ inch, $P = 10,000$ psi, and $S = 25,000$ psi?

3.2 A pressure vessel consists of an inner cylinder with an inside radius of 30 inches and a thickness of 3.5 inches and an outer cylinder with a thickness of 2.5 inches shrunk over the inner cylinder. During the shrinking process, strain gages on the inside surface of the inner cylinder measured an equivalent circumferential stress of 7000 psi in compression. Determine

1. The stress distribution in the inner and outer cylinders due to the shrinking process
2. The stress distribution in the inner and outer cylinders when an inside pressure of 4000 psi is applied

Let $E = 30,000$ ksi and $\mu = 0.3$.

3.3 In Example 3.5, assume a shearing force Q_o is applied at the end of the cylinder rather than M_o. Find the maximum longitudinal moment and its location from the edge. Let $\mu = 0.30$.

3.4 A cylindrical container is filled with a fluid to a level a–a. The metal temperature at a given time period is 400°F above section a–a and 100°F below section a–a. Determine the discontinuity forces in the cylinder at section a–a. Let $\alpha = 6.5 \times 10^{-6}$ inch/inch/°F, $E = 30,000$ ksi, and $\mu = 0.30$. Disregard forces due to fluid pressure.

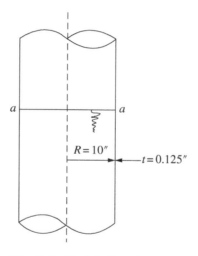

Problem 3.4   Cylindrical shell partially filled with fluid

3.5 Calculate the longitudinal bending stress at points a and b due to the applied loads shown. Let $E = 27,000$ ksi and $\mu = 0.32$.

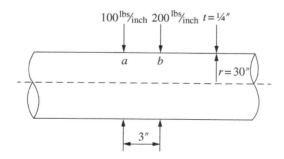

Problem 3.5   Multiple radial loads on a cylinder

3.6 Show that for a uniform load over a small length, a, the deflection at point A is given by

$$w = \frac{pr^2}{2Et}\left(2 - e^{-\beta c}\cos\beta c - e^{-\beta b}\cos\beta b\right).$$

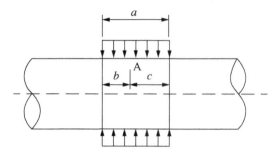

Problem 3.6   Distributed radial load on a cylinder

3.7 Find the expression for the bending moment in the water tank shown. Let $E = 20,000$ kgf/mm² and $\mu = 0.29$. *Hint*: Calculate first N_θ in terms of $p(x)$. Next use Eq. (3.72) to calculate w and dw/dx. Then use the fixed boundary condition at the bottom to calculate M_o. M_x is then obtained from Eq. (3.89).

How does M_o compare to that obtained by assuming a uniform pressure distribution at the junction rather than triangular?

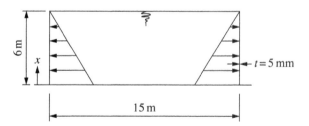

Problem 3.7   Storage tank

4

Buckling of Cylindrical Shells

4.1 Introduction

Cylindrical shells in boilers, pressure vessels, and nuclear components are generally designed for internal pressure as well as axial loads and external pressure. Tensile stresses due to internal pressure and axial forces are discussed in Chapters 1 through 3. Compressive stresses due to axial compressive loads and external pressure are discussed here.

Axial compressive forces on cylindrical shells cause either column-type buckling or shell buckling. In this chapter shell buckling is discussed.

4.2 Basic Equations

Cylindrical shells subjected to compressive forces, Figure 4.1, must be evaluated in accordance with one of the buckling theories of shells (Becker 1957, Gerard 1957a, 1957b, 1962). In this chapter Sturm's formulations (Sturm 1941) are presented. The formulation is well suited for designing cylindrical shells at various temperatures using actual stress–strain curves as discussed later. This approach is used in various ASME codes such as Section III and Section VIII-1.

Other formulations are also presented in this chapter. They include Miller's approach used in Section VIII-2 and the API approach used in parts of the nuclear code, Section III.

We begin Sturm's derivations by taking an infinitesimal element of a cylindrical shell with applied forces and moments as shown in Figure 4.2. The assumptions made in deriving the pertinent differential equations are as follows:

1. The cylinder is round before buckling.
2. The thickness is constant throughout the cylinder.
3. The material is isotropic, elastic, and homogeneous.

Stress in ASME Pressure Vessels, Boilers, and Nuclear Components, First Edition. Maan H. Jawad.
© 2018, The American Society of Mechanical Engineers (ASME), 2 Park Avenue,
New York, NY, 10016, USA (www.asme.org). Published 2018 by John Wiley & Sons, Inc.

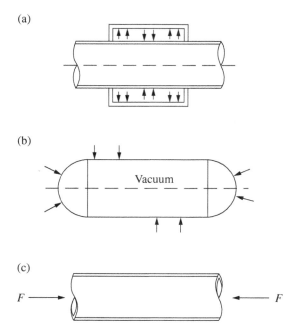

Figure 4.1 Compressive forces in cylindrical shells: (a) inside cylinder under lateral external pressure; (b) cylinder under external lateral and end pressure; and (c) cylinder under axial compressive force

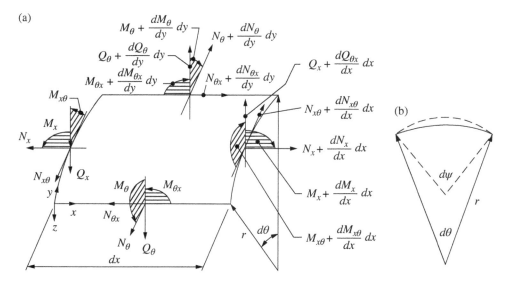

Figure 4.2 Infinitesimal element of a cylindrical shell: (a) forces and (b) deflection

4. The thickness is small compared to the radius.
5. Radial stress is negligible compared with the circumferential and longitudinal stresses.

Summation of forces and moments in the x-, y-, and z-directions results in the following six equations of equilibrium:

$$\frac{\partial N_x}{\partial x} + \frac{\partial N_{\theta x}}{\partial y} - Q_x \frac{\partial^2 w}{\partial x^2} = 0 \tag{4.1}$$

$$\frac{\partial N_\theta}{\partial y} + \frac{\partial N_{x\theta}}{\partial x} + Q_\theta \frac{\partial \psi}{\partial y} = 0 \tag{4.2}$$

$$\frac{\partial Q_\theta}{\partial y} + \frac{\partial Q_x}{\partial x} = p + N_\theta \frac{\partial \psi}{\partial y} - N_x \frac{\partial^2 w}{\partial x^2} - N_{x\theta} \frac{\partial^2 w}{\partial y \partial x} - N_{\theta x} \frac{\partial^2 w}{\partial x \partial y} \tag{4.3}$$

$$Q_x = \frac{\partial M_x}{\partial x} + \frac{\partial M_{\theta x}}{\partial y} \tag{4.4}$$

$$Q_\theta = \frac{\partial M_\theta}{\partial y} + \frac{\partial M_{x\theta}}{\partial x} \tag{4.5}$$

$$N_{\theta x} - N_{x\theta} + M_{\theta x} \frac{\partial \psi}{\partial y} + M_{x\theta} \frac{\partial^2 w}{\partial x^2} = 0 \tag{4.6}$$

where w is the deflection in the z-direction.

The previous equations cannot be solved directly because there are more unknowns than available equations. Accordingly, additional equations are needed. We can utilize the stress–strain relationships and rewrite them in terms of force–strain relationship as

$$N_x = \frac{Et}{1-\mu^2}(\varepsilon_x + \mu \varepsilon_\theta) \tag{4.7}$$

$$N_\theta = \frac{Et}{1-\mu^2}(\varepsilon_\theta + \mu \varepsilon_x) \tag{4.8}$$

$$N_{\theta x} = \frac{Et}{2(1+\mu)} \gamma_{\theta x} \tag{4.9}$$

where ε_x and ε_θ are the strains in the x- and y-directions and $\gamma_{\theta x}$ is the shearing strain.

Similarly, from Chapter 7 the moment–deflection equations are expressed as

$$M_x = -D\left(\frac{\partial^2 w}{\partial x^2} + \frac{\mu}{r^2}\frac{\partial^2 w}{\partial \theta^2} + \mu \frac{w}{r^2}\right) \tag{4.10}$$

$$M_\theta = -D\left(\mu \frac{\partial^2 w}{\partial x^2} + \frac{1}{r^2}\frac{\partial^2 w}{\partial \theta^2} + \frac{w}{r^2}\right) \tag{4.11}$$

$$M_{\theta x} = -D(1-\mu)\frac{1}{r}\frac{\partial^2 w}{\partial \theta \partial x}. \tag{4.12}$$

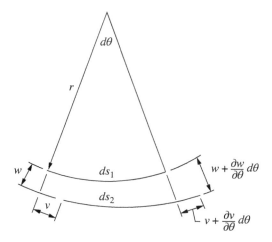

Figure 4.3   Deflected section

The relationships between strain and deformation are given by

$$\varepsilon_x = \frac{\partial \mu}{\partial x} \tag{4.13}$$

$$\varepsilon_\theta = \frac{\partial v}{\partial y} + \frac{v}{r} \tag{4.14}$$

$$\gamma_{\theta x} = \frac{\partial u}{\partial y} + \frac{\partial v}{\partial x} \tag{4.15}$$

where u and v are the deflections in the x- and y-directions, respectively.

Equation (4.14) is based on the fact that the radial deflection, w, for thin shells produces bending as well as stretching of the middle surface. Hence, from Figure 4.3,

$$\frac{ds_2 - ds_1}{ds_1} = \frac{(r+w)d\theta + (\partial v/\partial \theta)d\theta - rd\theta}{rd\theta}$$

$$= \frac{\partial v}{r\partial \theta} + \frac{w}{r} = \frac{\partial v}{\partial y} + \frac{w}{r}.$$

The change in angle $\delta\psi$ shown in Figure 4.2 is expressed as

$$\frac{\partial \psi}{\partial \theta} = 1 - \frac{1}{r}\frac{\partial^2 w}{\partial \theta^2} - \frac{w}{r} + \varepsilon_\theta. \tag{4.16}$$

This equation is obtained from Figure 4.2 where the angle $\delta\psi$ is the total sum of the following:

1. $d\theta$, which is the original angle
2. $-\dfrac{1}{r}\dfrac{\partial^2 w}{\partial \theta^2}d\theta$, which is the change in slope of length ds

3. $-\dfrac{w}{r}d\theta$, which is due to radial deflection

4. $\varepsilon_\theta\, d\theta$, which is due to circumferential strain

The derivative of these four expressions results in Eq. (4.16).

Assuming $M_{xy} = M_{yx}$, the previous 16 equations contain the following unknowns: N_x, N_θ, $N_{x\theta}$, $N_{\theta x}$, Q_x, Q_θ, M_x, M_θ, $M_{x\theta}$, ε_x, ε_θ, $\gamma_{x\theta}$, w, u, v, and ψ. These 16 equations can be reduced to four by the following various substitutions.

From Eqs. (4.13), (4.14), and (4.15), we get

$$\frac{\partial^2 \varepsilon_\theta}{\partial x^2} + \frac{\partial^2 \varepsilon_x}{\partial y^2} - \frac{\partial^2 \gamma_{\theta x}}{\partial x \partial y} - \frac{1}{r}\frac{\partial^2 w}{\partial x^2} = 0. \tag{4.17}$$

Substituting Eqs. (4.7), (4.8), and (4.9) into Eq. (4.17) gives

$$\frac{\partial^2 N_\theta}{\partial x^2} - \mu\frac{\partial^2 N_x}{\partial x^2} + \frac{1}{r^2}\frac{\partial^2 N_x}{\partial \theta^2} - \frac{\mu}{r^2}\frac{\partial^2 N_\theta}{\partial \theta^2} - \frac{2(1+\mu)}{r}\frac{\partial^2 N_{\theta x}}{\partial \theta \partial x} = \frac{Et}{r}\frac{\partial^2 w}{\partial x^2}. \tag{4.18}$$

The shearing forces in Eqs. (4.1) and (4.2) are eliminated by combining these two equations with Eqs. (4.4), (4.5), and (4.6). This gives

$$\frac{\partial^2 N_\theta}{\partial y^2} - \frac{\partial^2 N_x}{\partial x^2} + \frac{1+\varepsilon_\theta}{r}\frac{\partial^2 M_\theta}{\partial y^2} + \frac{2(1+\varepsilon_\theta)}{r}\frac{\partial^2 M_{\theta x}}{\partial y \partial x} = 0. \tag{4.19}$$

Equations (4.4) and (4.5) are combined with Eq. (4.3) to give

$$\frac{\partial^2 M_x}{\partial x^2} + 2\frac{\partial^2 M_{\theta x}}{\partial y \partial x} + \frac{\partial^2 M_\theta}{\partial y^2} = p + N_\theta\frac{\partial \psi}{\partial y} - N_x\frac{\partial^2 w}{\partial x^2} - N_{x\theta}\frac{\partial^2 w}{\partial y \partial x} - N_{\theta x}\frac{\partial^2 w}{\partial y \partial x}. \tag{4.20}$$

Substituting Eqs. (4.10), (4.11), and (4.12) into Eq. (4.20) results in the first of the four basic equations we are seeking:

$$-D\left[\frac{\partial^4 w}{\partial x^4} + \frac{\mu}{r^2}\frac{\partial^2 w}{\partial x^2} + \frac{2}{r^2}\frac{\partial^4 w}{\partial \theta^2 \partial x^2} + \frac{1}{r^4}\frac{\partial^4 w}{\partial \theta^4} + \frac{1}{r^4}\frac{\partial^2 w}{\partial \theta^2}\right]$$
$$= p + N_\theta\frac{1}{r}\frac{\partial \psi}{\partial \theta} - N_x\frac{\partial^2 w}{\partial x^2} - N_{x\theta}\frac{1}{r}\frac{\partial^2 w}{\partial \theta \partial x} - N_{\theta x}\frac{1}{r}\frac{\partial^2 w}{\partial \theta \partial x}. \tag{4.21}$$

The second basic equation is obtained by combining Eqs. (4.11), (4.12), and (4.19):

$$\frac{1}{r^2}\frac{\partial^2 N_\theta}{\partial \theta^2} - \frac{\partial^2 N_x}{\partial x^2} - \frac{(1+\varepsilon_\theta)}{r}D\left[\frac{1}{r^4}\frac{\partial^4 w}{\partial \theta^4} + \frac{1}{r^4}\frac{\partial^2 w}{\partial \theta^2} + \frac{(2-\mu)}{r^2}\frac{\partial^4 w}{\partial \theta^2 \partial x^2}\right] = 0. \tag{4.22}$$

Combining Eqs. (4.18) and (4.1) yields the third basic equation:

$$\frac{\partial^2 N_\theta}{\partial x^2} + (2+\mu)\frac{\partial^2 N_x}{\partial x^2} + \frac{1}{r^2}\frac{\partial^2 N_x}{\partial \theta^2} - \frac{\mu}{r^2}\frac{\partial^2 N_\theta}{\partial \theta^2} = \frac{Et}{r}\frac{\partial^2 w}{\partial x^2}. \tag{4.23}$$

Solving for ε_θ from Eqs. (4.7) and (4.8) and differentiating twice with respect to x gives the fourth basic equation:

$$\frac{\partial^2 N_\theta}{\partial x^2} - \mu\frac{\partial^2 N_x}{\partial x^2} = \frac{Et}{r}\left(\frac{\partial^3 v}{\partial x^2 \partial \theta} + \frac{\partial^2 w}{\partial x^2}\right). \tag{4.24}$$

Equations (4.21) through (4.24) are the four basic equations needed to develop a solution for the buckling of cylindrical shells.

4.3   Lateral Pressure

If the external pressure is applied only to the side of the cylinder as shown in Figure 4.1a, then the solution can be obtained as follows. Let

$$N_\theta = -pr + f(x,y) \tag{4.25}$$

where $f(x, y)$ is a function of x and y that expresses the variation of N_θ from the average value. If the deflection, w, is small, then the function $f(x, y)$ is also very small. Similarly, the end force N_x is expressed as

$$N_x = 0 + g(x,y). \tag{4.26}$$

Since $N_{\theta x} = N_{x\theta} = 0$, then

$$N_{\theta x} = 0 + h(x,y) \tag{4.27}$$

$$N_{x\theta} = 0 + j(x,y). \tag{4.28}$$

Substituting Eqs. (4.25) through (4.28) into Eqs. (4.21) through (4.24) and neglecting higher order terms such as

$$g\frac{\partial^2 w}{\partial \theta \partial x}, h\frac{\partial^2 w}{\partial \theta \partial x}, j\frac{\partial^2 w}{\partial \theta \partial x}, \quad \text{and } f \text{ with terms in } \frac{\partial \psi}{\partial \theta}$$

other than unity result in the following four equations:

$$D\left[\frac{\partial^4 w}{\partial x^4} + \frac{\mu}{r^2}\frac{\partial^2 w}{\partial x^2} + \frac{2}{r^2}\frac{\partial^4 w}{\partial \theta^2 \partial x^2} + \frac{1}{r^4}\left(\frac{\partial^2 w}{\partial \theta^2} + \frac{\partial^4 w}{\partial \theta^4}\right)\right] + \frac{1}{r}f(x,y)$$

$$= -p\left[\frac{1}{r}\frac{\partial^2 w}{\partial \theta^2} + \frac{w}{r} + \varepsilon_\theta\right] \tag{4.29}$$

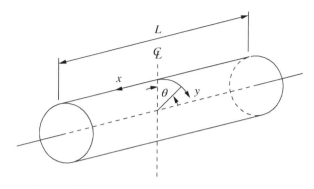

Figure 4.4 Coordinate system in a cylindrical shell

$$\frac{1}{r^2}\frac{\partial^2 f}{\partial\theta^2} - \frac{\partial^2 g}{\partial x^2} - \frac{(1+\varepsilon_\theta)}{r^3}D\left[\frac{1}{r^2}\frac{\partial^4 w}{\partial\theta^4} + \frac{1}{r^2}\frac{\partial^2 w}{\partial\theta^2} + (2-\mu)\frac{\partial^4 w}{\partial\theta^2\partial x^2}\right] = 0 \tag{4.30}$$

$$\frac{\partial^2 f}{\partial x^2} + (2+\mu)\frac{\partial^2 g}{\partial x^2} + \frac{1}{r^2}\frac{\partial^2 g}{\partial\theta^2} - \frac{\mu}{r^2}\frac{\partial^2 f}{\partial\theta^2} = \frac{Et}{r}\frac{\partial^2 w}{\partial x^2} \tag{4.31}$$

$$\frac{\partial^2 f}{\partial x^2} - \mu\frac{\partial^2 g}{\partial x^2} = \frac{Et}{r}\left(\frac{\partial^3 v}{\partial x^2\partial\theta} + \frac{\partial^2 w}{\partial x^2}\right). \tag{4.32}$$

Equations (4.29) through (4.32) can be solved for various boundary conditions. For a simply supported cylinder, the following conditions are obtained from Figure 4.4:

$$\text{at } x = \pm\frac{L}{2}, w = \frac{\partial^2 w}{\partial x^2} = \frac{\partial^2 w}{\partial\theta^2} = 0.$$

Also, because of symmetry,

$$\frac{\partial w}{\partial\theta} = 0 \text{ for all values of } \theta \text{ when } x = 0$$

and

$$\frac{\partial w}{\partial\theta} = 0 \text{ for all values of } x \text{ when } \theta = 0.$$

Similarly, $\partial v/\partial\theta = 0$ for all values of θ at $x = \pm L/2$.

These boundary conditions suggest a solution of the form

$$w = A\cos n\theta\cos\frac{\pi x}{L}$$

$$v = B\sin n\theta\cos\frac{\pi x}{L}$$

where n is the number of lobes as defined in Figure 4.5. Substituting these two expressions into Eqs. (4.29) through (4.32) gives

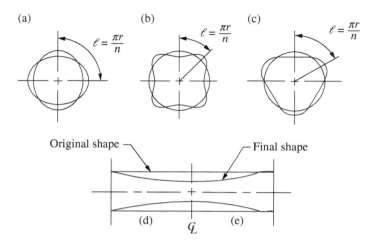

Figure 4.5 Buckling modes of a cylindrical shell: (a) $n = 2$, (b) $n = 4$, (c) $n = 3$, (d) edges simply supported symmetrical about $\mathbb{C}$, and (e) edges fixed symmetrical about $\mathbb{C}$

$$D\left[\frac{\pi^4}{L^4} + \frac{(2n^2 - \mu)\pi^2}{r^2 L^2} - \frac{n^2}{r^4} + \frac{n^4}{r^4}\right] A\cos n\theta \cos\frac{\pi x}{L}$$

$$= \frac{p}{r}(An^2 + Bn)\cos n\theta\cos\frac{\pi x}{L} - \frac{1}{r}f(x,y)$$
(4.33)

$$\frac{1}{r^2}\frac{\partial^2 f}{\partial\theta^2} - \frac{\partial^2 g}{\partial x^2} = \frac{(1-\varepsilon_\theta)}{r^3}D\left[\frac{n^4}{r^2} - \frac{n^2}{r^2} + (2-\mu)\frac{n^2\pi^2}{L^2}\right]A\cos n\theta\cos\frac{\pi x}{L}$$
(4.34)

$$(2+\mu)\frac{\partial^2 g}{\partial x^2} + \frac{1}{r^2}\frac{\partial^2 g}{\partial\theta^2} + \frac{\partial^2 f}{\partial x^2} - \frac{\mu}{r^2}\frac{\partial^2 f}{\partial\theta^2} = \frac{-Et\,\pi^2}{r\,L^2}A\cos n\theta\cos\frac{\pi x}{L}$$
(4.35)

$$\frac{\partial^2 f}{\partial x^2} - \mu\frac{\partial^2 g}{\partial x^2} = -\frac{Et\,\pi^2}{r\,L^2}(Bn + A)\cos n\theta\cos\frac{\pi x}{L}.$$
(4.36)

From Eq. (4.33) it follows that

$$f(x,y) = C\cos n\theta\cos\frac{\pi x}{L}$$
(4.37)

and from Eqs. (4.34) and (4.35) that

$$g(x,y) = G\cos n\theta\cos\frac{\pi x}{L}.$$
(4.38)

From Eq. (4.36)

$$Bn + A = \frac{r}{Et}(C - \mu G)$$
(4.39)

and the values of C and G are found to be

$$\frac{C}{A} = \frac{Et}{r\alpha^2} - \frac{Et}{r^3(1-\mu^2)}(H-1)\left(1-\frac{pr}{Et}\right)\left(\frac{\alpha+1-\mu}{\lambda\alpha}\right) \tag{4.40}$$

$$\frac{G}{A} = \frac{Et}{r\alpha\lambda} + \frac{Et}{r^3(1-\mu^2)}(H-1)\left(1-\frac{pr}{Et}\right)\left(\frac{1-\mu(\alpha-1)}{\lambda\alpha}\right) \tag{4.41}$$

where

$$H = n^2[1 + (\lambda-1)(2-\mu)]$$

$$\lambda = \frac{\pi^2 r^2}{n^2 L^2} + 1$$

$$\alpha = \frac{n^2 L^2}{\pi^2 r^2} + 1.$$

From Eq. (4.33),

$$\frac{Et}{r^2\alpha^2} - \frac{D}{r^4}\frac{\alpha+1+\mu}{\alpha\lambda}(H-1)A\cos n\theta\cos\frac{\pi x}{L}$$

$$+ \frac{D}{r^4}n^2\left(n^2\lambda^2 - \mu(\lambda-1) - 1\right)A\cos n\theta\cos\frac{\pi x}{L} \tag{4.42}$$

$$= \frac{p}{r}FA\cos n\theta\cos\frac{\pi x}{L}$$

where

$$F = n^2 - 1 + \frac{1}{\alpha} - \frac{\mu}{\alpha\lambda} - \frac{D}{r^2 Et\alpha\lambda}$$

$$\times \left\{(H-1)\left(1-\frac{pr}{Et}\right)\left[\alpha(1-\mu^2) + (1+\mu^2)\right] + \alpha + 1 + \mu\right\}. \tag{4.43}$$

Equation (4.42) indicates that solutions different from zero exist only if

$$p_{cr} = \frac{X+Y-Z}{F} \tag{4.44}$$

where

$$X = \frac{Et}{r\alpha^2}$$

$$Y = \frac{D}{r^3}n^2\left[n^2\lambda^2 - \mu(\lambda-1) - 1\right]$$

$$Z = \frac{D}{r^3}\frac{\alpha+1+\mu}{\alpha\lambda}(H-1).$$

Equation (4.44) can be written as

$$p_{cr} = \frac{1}{8}KE\left(\frac{t}{r}\right)^3 \tag{4.45}$$

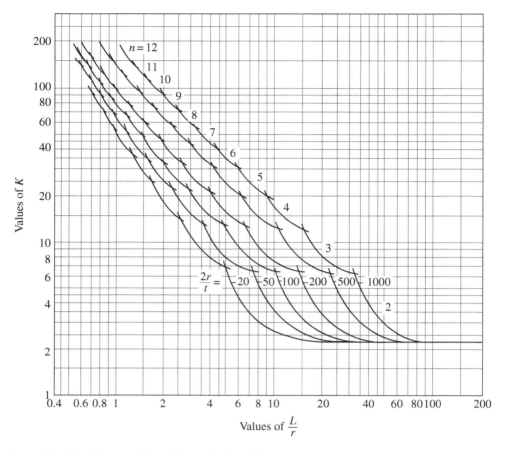

Figure 4.6 Buckling coefficient K for cylinders with pressure on sides only, edges simply supported; $\mu = 0.30$. Source: Adapted from Sturm (1941)

where

$$K = K_1 + 4K_2 \left(\frac{r}{t}\right)^2 \tag{4.46}$$

$$K_1 = \frac{2n^2}{3} \frac{\left[n^2\lambda^2 - \mu(\lambda - 1)\right] - U(H - 1)}{F(1 - \mu^2)} \tag{4.47}$$

$$K_2 = \frac{2}{\alpha^2 F} \tag{4.48}$$

and

$$U = \frac{\alpha + 1 + \mu}{\alpha\lambda}.$$

Equation (4.45) is the basic equation for the buckling of cylindrical shells subjected to lateral pressure. A plot of K in Eq. (4.45) is shown in Figure 4.6.

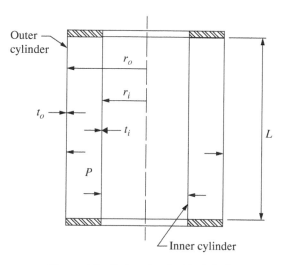

Figure 4.7 Jacketed cylindrical shell

Example 4.1

Find the allowable external pressure for the inner cylinder shown in Figure 4.7. Let $L = 10$ ft, $r_i = 2$ ft, $E = 29{,}000$ ksi, $t_i = 1/2$ inch, factor of safety (F.S.) $= 2.5$, and $\mu = 0.3$. Assume the inner cylinder to be simply supported.

Solution

$$\frac{L}{r} = 5, \quad \frac{2r}{t} = 96$$

From Figure 4.6, $K = 11$. Hence, from Eq. (4.45),

$$P_{cr} = \frac{11}{8}(29{,}000{,}000)\left(\frac{0.5}{24}\right)^3$$

$$= 360\,\text{psi}$$

$$p = \frac{P_{cr}}{\text{F.S.}} = \frac{360}{2.5} = 144\,\text{psi}.$$

Equation (4.45) assumes the end of the cylindrical shell to be simply supported. A similar equation can be derived for the case of a cylindrical shell with fixed ends. In this case the slope and deflection at the ends are zero. Proceeding in a similar fashion as for the simply supported case, a buckling equation is obtained. The derivation is more complicated than that for the simply supported cylinder. The resulting buckling equation is the same as Eq. (4.45) with the exception of the value of K. A plot of K for the fixed-end condition is shown in Figure 4.8.

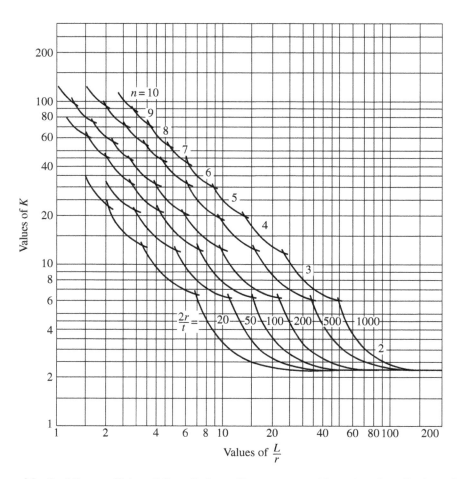

Figure 4.8 Buckling coefficient K for cylinders with pressure on sides only, edges fixed; $\mu = 0.30$. Source: Adapted from Sturm (1941)

4.4 Lateral and End Pressure

Many cylindrical shells are subjected to axial forces in the lateral and axial directions, Figure 4.1b, or to vacuum. The governing equations are very similar to those derived for the lateral condition. For lateral and axial loads, Eq. (4.26) is written as

$$N_x = -\frac{pr}{2} + g(x,y) \tag{4.49}$$

while Eqs. (4.25), (4.27), and (4.28) remain the same. Equation (4.29) becomes

$$D\left[\frac{\partial^4 w}{\partial x^4} + \frac{\mu}{r}\frac{\partial^4 w}{\partial x^4} + \frac{2}{r^2}\frac{\partial^4 w}{\partial \theta^2 \partial x^2} + \frac{1}{r^4}\frac{\partial^2 w}{\partial \theta^2} + \frac{\partial^4 w}{\partial \theta^4}\right] + \frac{1}{r}f(x,y)$$

$$= -p\left[\frac{1}{r}\frac{\partial^2 w}{\partial \theta^2} + \frac{w}{r} + \varepsilon_\theta\right] - \frac{pr}{2}\frac{\partial^2 w}{\partial x^2} \tag{4.50}$$

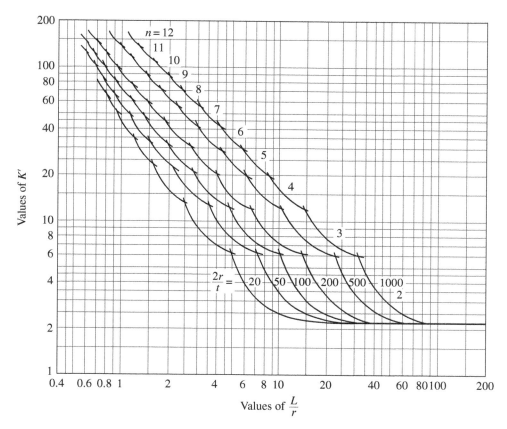

Figure 4.9 Buckling coefficient K' for cylinders with pressure on sides and ends, edges simply supported; $\mu = 0.30$. Source: Adapted from Sturm (1941)

while Eqs. (4.30), (4.31), and (4.32) remain the same. Using the boundary conditions for a simply supported cylinder, the governing Eq. (4.45) can be written as

$$p_{\mathrm{cr}} = \frac{1}{8} K' E \left(\frac{t}{r}\right)^3 \tag{4.51}$$

where

$$K' = K'_1 + 4K'_2 \left(\frac{r}{t}\right)^2$$

$$K'_1 = K_1 \frac{F}{F + (\pi^2 r^2 / 2L^2)} \tag{4.52}$$

$$K'_2 = K_2 \frac{F}{F + (\pi^2 r^2 / 2L^2)}.$$

A plot of Eq. (4.52) for a simply supported cylinder is shown in Figure 4.9.

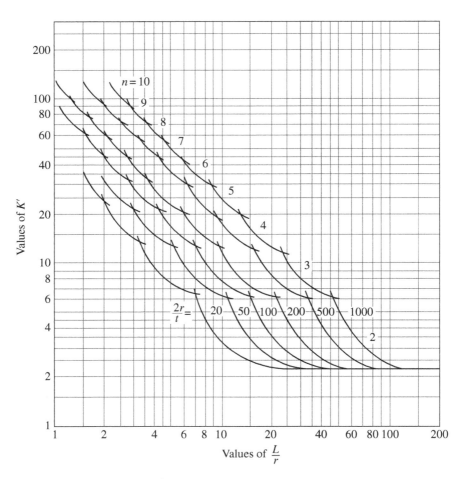

Figure 4.10 Buckling coefficient K' for cylinders with pressure on sides and ends, edges fixed; $\mu = 0.30$. Source: Adapted from Sturm (1941)

A similar equation can be derived for a cylinder with fixed ends. The resulting K' value is plotted in Figure 4.10.

Example 4.2

Determine if the fuel tank shown in Figure 4.11 is adequate for full vacuum condition. Let $E = 16,000$ ksi, $\mu = 0.3$ and factor of safety (F.S.) = 4.

Solution

$$\frac{L}{r} = \frac{96}{24} = 4.00$$

$$\frac{2r}{t} = 192.$$

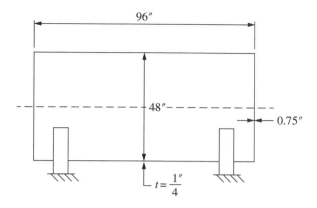

Figure 4.11   Gasoline tank

Assume the ends of the cylinder to be simply supported due to the flexibility of the end plates. From Figure 4.9, $K' = 20$ and from Eq. (4.51),

$$p = \frac{1}{4}\left[\frac{20}{8} \times 16,000,000\left(\frac{0.25}{24}\right)^3\right]$$

$$= 11.3\,\text{psi} < 14.7\,\text{psi. no good.}$$

For $t = 5/16$ inch, we get $K' = 16$ and $p = 17.66$ psi.

4.5   Axial Compression

When an axial force is applied at the end of a cylinder, the buckling strength is developed as follows. Let W be the applied end load per unit length of circumference. Equations (4.25) and (4.26) become

$$N_\theta = 0 + f(x, y) \tag{4.53}$$

$$N_x = -W + g(x, y). \tag{4.54}$$

For a simply supported cylinder, the buckling equation can be expressed as

$$W_{cr} = \frac{X + Y - Z}{n^2 + V(H_0 - 1)} \tag{4.55}$$

where

$$X = \frac{Et}{\alpha_o \lambda_o}$$

$$Y = \frac{D(\alpha_o - 1)}{r^2}\left\{n^2\left[n^2\lambda_0^2 - \mu(\lambda_o - 1)\right] - 1\right\}$$

$$Z = \frac{D(a_0 - 1)}{r^2} \frac{a_0 + 1 + \mu}{a_0 \lambda_0}(H_0 - 1)$$

$$V = \frac{\mu D(a_0 + 1 + \mu)}{r E t^2 \lambda_0^2}$$

$$H_0 = n^2 \left[i + (\lambda_0 - 1)(2 - \mu) \right]$$

$$a_0 = \frac{n^2 L_0^2}{\pi^2 r^2} + 1$$

$$\lambda_0 = \frac{\pi^2 r^2}{n^2 L_0^2} + 1$$

L_0 = length of one buckle wave $(L_0 << L)$.

Equation (4.55) can be simplified for design purposes.

Let $\sigma_{cr} = W_{cr}/t$, and for long cylinders with large values of r/t, Eq. (4.55) becomes

$$\sigma_{cr} = \frac{0.6Et}{r}. \tag{4.56}$$

Tests have shown that for actual cylinders, which have a slight out-of-roundness, the stress magnitude given by Eq. (4.56) tends to be unconservative for large r/t ratios. Figure 4.12 shows the scatter of some of the data obtained from various tests. An empirical equation that defines the lower bound of available test data (ACI 1981) is given by

$$\sigma_{cr} = \frac{0.6CEt}{r} \tag{4.57}$$

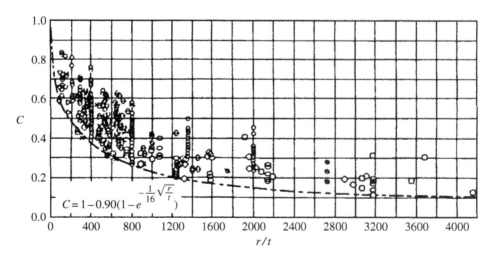

Figure 4.12 Comparison of Eq. (4.56) with test data. Source: Seide (1981). Reproduced with permission of ACI

where

$$C = 1 - 0.9\left(1 - e^{(-1/16)\sqrt{(r/t)}}\right).$$

Example 4.3
Determine the thickness of the lower cylinder shown in Figure 4.13 needed to support a reactor weighing 400,000 lbs. Use Eq. (4.57). Let $E = 30,000$ ksi and factor of safety (F.S.) $= 4.0$.

Solution
Try $t = 0.25$ inch. Then $r/t = 192$.
From Eq. (4.57), $C = 0.48$
and

$$\sigma = \frac{1}{4}(0.48 \times 0.6 \times 30,000,000 \times 0.25/48)$$

$$= 11,250\,\text{psi}.$$

Actual stress is given by

$$= \frac{400,000}{\pi \times 96 \times 0.25} = 5300\,\text{psi}.$$

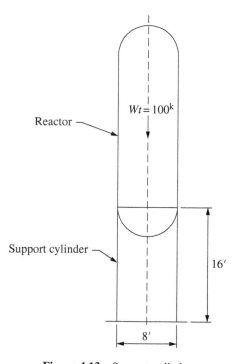

Figure 4.13   Support cylinder

Theoretically, the thickness can be reduced further. However, consideration must be given to handling and field erection procedures.

It should be noted that the buckling equations derived for lateral pressure, lateral plus axial pressure, and axial forces are based on numerous assumptions and approximations. Experimental tests have shown that these assumptions and approximations tend to make the calculated critical buckling pressure unconservative for large r/t ratios. This is partly due to out-of-roundness of large r/t cylinders caused by such factors as fabrication forming processes, weld shrinkage, and installation tolerances. Accordingly, large factors to safety are usually applied to the calculated buckling pressure when used for design purposes. This is explained further in Section 4.6 under design equations.

4.6 Design Equations

The equations derived in Sections 4.1 through 4.5 are used in numerous codes and standards for design purposes. Some of these design equations are given in this section.

4.6.1 External Pressure

The ASME Sections VIII-I and Sections III-NC and ND use Eq. (4.51) as a basis for establishing design rules for external loads. This method permits the use of stress–strain curves of actual materials of construction to obtain allowable external pressure. This procedure prevents the possibility of calculating an allowable external pressure that results in a stress value that is above the yield stress of the material. Equation (4.51) for lateral and end pressure can be written as

$$\sigma_{\text{cr}} = \frac{p_{\text{cr}}r}{t} = \frac{K'}{2}E\left(\frac{t}{2r}\right)^2. \tag{4.58}$$

Define

$$\varepsilon_{\text{cr}} = \frac{\sigma_{\text{cr}}}{E}.$$

Equation (4.58) becomes

$$\varepsilon_{\text{cr}} = \frac{K'}{2}\left(\frac{t}{2r}\right)^2. \tag{4.59}$$

A plot of this equation as a function of strain (designated as "A" in the ASME code), L/D_o, and D_o/t, where D_o is the outside diameter of the shell, is shown in Figure 4.14. This geometric figure is used by entering the values of L/D_o and t/D_o for a given shell and determining the strain A.

In order to determine the allowable external pressure, plots of stress–strain diagrams are used for different materials at various temperatures. From these diagrams plots of strain versus

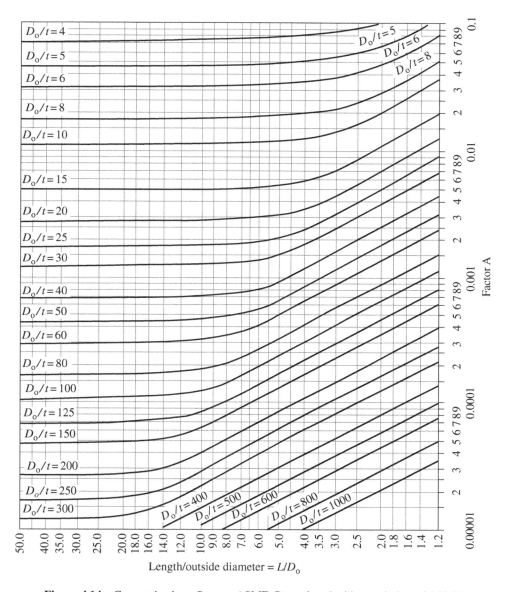

Figure 4.14 Geometric chart. Source: ASME. Reproduced with permission of ASME

elastic moduli and tangent moduli are developed and plotted on a log-log scale. The justification for these plots is explained in Section 4.6.2. The ordinate in these plots is designated as stress B and the abscissa as strain A. The values of B are further limited to one-half the yield stress of the material at a given temperature. A sample of such external pressure chart is shown

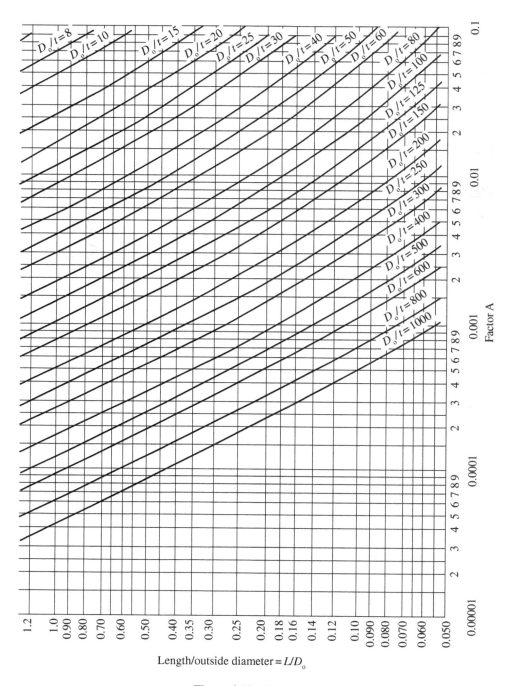

Figure 4.14   (Continued)

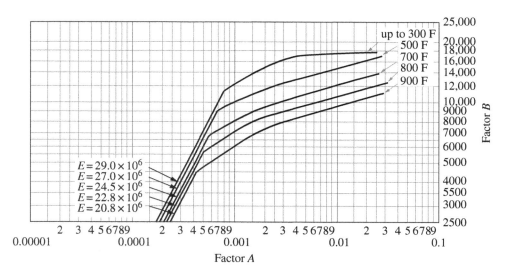

Figure 4.15 External pressure chart for carbon and low alloy steels. Source: ASME. Reproduced with permission of ASME

in Figure 4.15. The ASME procedure consists of determining strain A from Figure 4.14 and then B value from Figure 4.15. The allowable external pressure is then calculated from

$$P = \frac{(2B)t}{(F.S.)r} = \frac{4B}{(F.S.)(D_o/t)}. \tag{4.60}$$

If the value of A falls to the left of the material curves in Figure 4.15, then the allowable external pressure is given by

$$P = \frac{AE}{(F.S.)(r/t)} = \frac{2AE}{(F.S.)(D_o/t)}. \tag{4.61}$$

The ASME code, Sections III-NB, III-NC, III-ND, and VIII-1, use a factor of safety (F.S.) of 3.0 in Eqs. (4.60) and (4.61) for external pressure design. These equations then become

$$P = \frac{4B}{3(D_o/t)}. \tag{4.62}$$

$$P = \frac{2AE}{3(D_o/t)}. \tag{4.63}$$

The ability of a cylindrical shell to resist external pressure increases with a reduction in its effective length. The ASME code gives rules for adding stiffening rings to reduce the effective length. Rules for the design of stiffening rings (Jawad and Farr 1989) are also given in the ASME code.

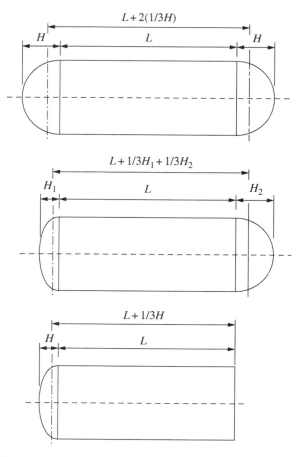

Figure 4.16 Cylindrical shells with various end attachments

An empirical equation developed by the US Navy (Raetz 1957) for the buckling of cylindrical shells under lateral and axial pressure in the elastic range is given by

$$p_{cr} = \frac{2.42E}{(1-\mu^2)^{3/4}} \frac{(t/2r)^{2.5}}{\left[L/2r - 0.45(t/2r)^{1/2}\right]}. \tag{4.64}$$

This equation is a good approximation of Eq. (4.51).

Tests that led to the development of Eq. (4.64) also showed that the effective cylindrical length for cylinders with end closures in the form of hemispherical or elliptical shape is equal to the length of the actual cylinder plus one-third the depth of the end closures as illustrated in Figure 4.16.

For structures with large r/t ratios, Eq. (4.64) can be simplified to

$$p_{cr} = \frac{2.42E}{(1-\mu^2)^{3/4}} \frac{(t/D_o)^{2.5}}{L/D_o}.$$

Substituting into this equation the value of $\mu = 0.3$ for metallic structures results in

$$p_{cr} = \frac{2.60E(t/D_o)^{2.5}}{L/D_o}. \tag{4.65}$$

Example 4.4

Determine the allowable pressure for the cylinder in Figure 4.17 based on Eq. (4.65). Compare the result with that obtained from ASME's Eq. (4.62). Let $E = 29,000$ ksi and factor of safety (F.S.) = 3.0.

Solution

$$\frac{L}{D_o} = \frac{(96 + 2 \times 9/3)}{48} = 2.13$$

$$\frac{t}{D_o} = 0.0052$$

From Eq. (4.65),

$$P = \frac{2.60(29,000,000)(0.0052)^{2.5}}{(3.0)(2.13)} = 23.0 \, \text{psi.}$$

Using the ASME method, $D_o/t = 192$.
From Figure 4.14, $A = 0.00024$.
From Figure 4.15, $B = 3400$ psi.
From Eq. (4.62),

$$P = \frac{4(3400)}{(3)(192)} = 23.6 \, \text{psi.}$$

Another methodology for calculating allowable external pressure in a cylinder was developed by Miller (1999) and is used in Section VIII-2 of the ASME code. Miller combined

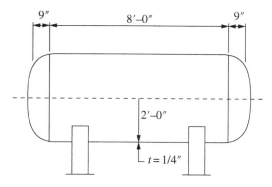

Figure 4.17   Storage vessel

experimental work he performed on flat bottom tanks with theoretical work and developed the following procedure:

1. Assume a thickness t for a cylinder with a given outside diameter D_o, outside radius R_o, and effective length L.
2. Calculate elastic buckling F_{he} as follows:

$$M_x = \frac{L}{(R_o t)^{0.5}}$$

$$C_h = 0.55 \left(\frac{t}{D_o}\right) \quad \text{for} \quad M_x \geq 2\left(\frac{D_o}{t}\right)^{0.94}$$

$$C_h = 1.12 M_x^{-1.058} \quad \text{for} \quad 13 < M_x < 2\left(\frac{D_o}{t}\right)^{0.94}$$

$$C_h = 1.0 \quad \text{for} \quad 1.5 < M_x \leq 13$$

$$F_{he} = \frac{1.6 C_h E_y t}{D_o}$$

3. Calculate the predicted buckling stress F_{1c} as follows:

$$F_{1c} = S_y \quad \text{for} \quad F_{he}/S_y \geq 2.439$$

$$F_{1c} = 0.7 S_y \left(F_{he}/S_y\right)^{0.4} \quad \text{for} \quad 0.552 < F_{he}/S_y < 2.439$$

$$F_{1c} = F_{he} \quad \text{for} \quad F_{he}/S_y < 0.552$$

4. Calculate the design factor F.S. as follows:

$$\text{F.S.} = 2.0 \quad \text{for} \quad F_{1c} \leq 0.55 S_y$$

$$\text{F.S.} = 2.407 - 0.741 \left(\frac{F_{1c}}{S_y}\right) \quad \text{for} \quad 0.55 S_y < F_{1c} < S_y$$

$$\text{F.S.} = 1.667 \quad \text{for} \quad F_{1c} = S_y$$

5. Calculate the allowable external pressure P_a as follows:

$$F_{ha} = \frac{F_{1c}}{\text{F.S.}}$$

$$P_a = 2 F_{ha} \left(\frac{t}{D_o}\right). \tag{4.66}$$

Example 4.5

What is the allowable external pressure of the vessel shown in Figure 4.17? Use Eq. (4.66) and let $E = 29{,}000{,}000$ psi and $S_y = 38{,}000$ psi.

Solution

$$M_x = \frac{102}{[(24)(1/4)]^{0.5}} = 41.64$$

$$C_h = 1.12(41.64)^{-1.058} = 0.0217$$

$$F_{he} = 1.6(0.0217)(29{,}000{,}000)(0.25)/48 = 5244 \text{ psi}$$

$$\text{F.S.} = 2.0$$

$$F_{ha} = \frac{5244}{2} = 2622 \text{ psi}$$

$$P_a = 2(2622)(0.25)/48 = 27.3 \text{ psi}$$

This allowable external pressure is about 15% higher than that of Example 4.4 using VIII-1 methodology.

4.6.2 Axial Compression

4.6.2.1 Elastic Buckling

The critical elastic axial compression stress for a long cylindrical shell is obtained from Eq. (4.56) as

$$\sigma_{cr} = \frac{0.6E}{(R_o/t)}. \tag{4.67}$$

The critical strain is then given by

$$\varepsilon_{cr} = \frac{0.6}{(R_o/t)} \tag{4.68}$$

where

$$\varepsilon_{cr} = \sigma_{cr}E. \tag{4.69}$$

Experimental data shows that buckling for cylinders with large R_o/t may occur at a value of 1/10th that given by the previous equation. Accordingly ASME uses a knockdown factor of 5 in Eq. (4.68) to account for geometric imperfections and a design factor of 2.0 in Eq. (4.69) to account for variations in material properties and elastic modulus. Equations (4.68) and (4.69) then become

$$A = \frac{0.125}{(R_o/t)} \tag{4.70}$$

and

$$B = \frac{AE}{2} \quad \text{but not greater than allowable tensile stress} \tag{4.71}$$

where
A = a nondimensional factor based on geometry
B = allowable compressive stress
E = Young's modulus of elasticity

4.6.2.2 Inelastic Buckling

Equation (4.67) in only applicable in the elastic region where E is constant. In the inelastic region, the buckling equation is a function of the tangent and secant moduli and is given by

$$\sigma_{cr} = \frac{[E_s E_t / 3 (1 - v^2)]^{0.5}}{(R_o / t)} \tag{4.72}$$

where
E = Young's modulus of elasticity
E_s = secant modulus
E_t = tangent modulus
μ = Poisson's ratio
$v = 0.5 - (0.5 - \mu)(E_s / E)$

In the elastic range, $E_s = E_t = E$ and Eq. (4.72) becomes the same as Eq. (4.68). In the inelastic range, the value of E_s can conservatively be assumed to be the same as E_t in order to simplify the application of Eq. (4.72). ASME then uses this equation to develop design charts beyond the elastic range. This is done by using stress–strain diagrams for a given material at a given temperature. From these diagrams, a chart is developed correlating the tangent modulus versus A in the inelastic range and bound by an elastic line correlating Young's modulus to A. This chart is referred to by ASME as an external pressure chart. One such chart is shown in Figure 4.15.

Example 4.6
The reactor shown in Figure 4.18a is constructed with carbon steel with yield stress of 38 ksi and E = 29,000 ksi and allowable stress, S = 20 ksi. The design temperature is 100°F. (a) Determine by the ASME method the required thickness due to vacuum. (b) Check the thickness of the vessel due to the wind loading shown in Figure 4.18b.

Solution
a. Vacuum condition

$$\text{Try } t = \frac{3}{8} \text{ inch.}$$

$$L = 30 + (2)(1)/3 = 30.67 \, \text{ft} = 368 \text{ inch.}$$

$$\frac{L}{D_o} = 7.7 \text{ and } \frac{D_o}{t} = 128.$$

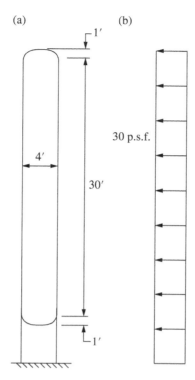

(a)   (b)

30 p.s.f.

4′

30′

Figure 4.18   Reactor vessel: (a) vertical vessel and (b) wind load

From Figure 4.14, $A = 0.00012$ inch/inch.
From Figure 4.15, A falls to the left of the material curve.
Hence, from Eq. (4.63)

$$p = \frac{2(0.00012)(29,000,000)}{3(128)} = 18.1\,\text{psi (acceptable stress)}.$$

b. The bending moment at the bottom of the cylinder due to wind load is given by

$$M = (30)(4)(31)^2/2 = 57\ 660\ \text{ft-lbs}$$

$$\text{stress}\ \sigma = \frac{M}{\pi r^2 t} = \frac{57,660 \times 12}{\pi (24)^2 (0.375)} = 1020\ \text{psi}.$$

From Eq. (4.70),

$$A = \frac{0.125}{24/0.375} = 0.002\ \text{inch/inch}$$

and from Figure 4.15, the allowable compressive stress is determined as

$$B = 15,000\,\text{psi} > 1020\,\text{psi} \ (\text{acceptable stress}).$$

Check Euler's buckling $(K = 2.0)$ by calculating

$$\text{Area} = 2\pi rt = 56.55 \ \text{inch}^2 \quad \text{and} \quad I = \pi r^3 t = 16,286 \ \text{inch}^4$$

$$\sigma_{cr} = \frac{\pi^2 \times 29,000,000 \times 16,286}{(2 \times 360)^2 (56.55)}$$

$$\sigma_{cr} = 159,000 \ \text{psi}.$$

Use 20,000 psi allowable stress.

$$\text{F.S.} = \frac{20,000}{1020} = 20.$$

Hence, Euler's buckling does not control.

An empirical equation that is often used in the design of stacks and other self-supporting cylindrical structures made of low-carbon steel was developed by the Chicago Bridge and Iron Company (Roark and Young 1975) as

$$\text{Allowable stress} = (X)(Y) \tag{4.73}$$

where

$$X = [1,000,000 \ t/r] \left[2 - \frac{2}{3}(100 \ t/r) \right] \quad \text{for} \ t/r < 0.015$$

$$X = 15,000 \ \text{psi} \ \text{for} \ t/r > 0.015$$

$$Y = 1 \ \text{for} \ 2L/r \leq 60$$

$$Y = \frac{21,600}{18,000 + 2(L/r)} \quad \text{for} \ 2L/r > 60.$$

Minimum $t = 1/4$ inch.

Another expression that is often used in determining the allowable compressive stress of steel cylindrical shells of flat bottom oil storage tanks was developed by the API (API Standard 620) and used in the ASME nuclear code Section III, subsection ND. From Eq. (4.66) with $E = 30,000,000$ psi and a factor of safety of 10, the allowable compressive stress becomes

$$\sigma = 1,800,000 \left(\frac{t}{r}\right). \tag{4.74}$$

This allowable compressive stress is limited to 15,000 psi. A stress transition equation between this limit and Eq. (4.67) was developed by API as shown in Figure 4.19. The API standard also includes curves for allowable stress of cylindrical shells subjected to biaxial stress combinations.

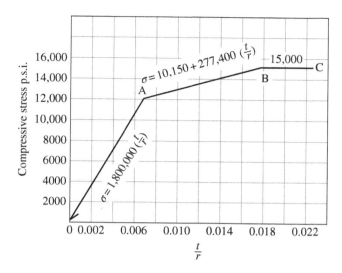

Figure 4.19 Allowable compressive stress in flat bottom tanks. Source: ASME. Reproduced with permission of ASME

It should be noted that Eqs. (4.67) and (4.74) do not include any terms for the length of the cylinder. Hence, for extremely large L/r ratios, Euler's equation for column buckling may control the allowable stress rather than shell buckling and should be checked. In this case, the expression to be considered is

$$\sigma_{\text{cr}} = \frac{\pi^2 EI}{(KL)^2 A} \tag{4.75}$$

where
A = cross-sectional area of cylinder
I = moment of inertia of cylinder
K = 1.0 for cylinders with simply supported ends
= 0.5 for cylinders with fixed ends
= 2.0 for cantilever cylinders

Example 4.7
The stack shown in Figure 4.20 is subjected to an effective wind pressure of 40 psf. What is the required thickness? Use Eq. (4.73).

Solution
Maximum bending moment at the bottom

$$M = (40)(8)(50)^2/2 = 400,000 \, \text{ft-lbs.}$$

Try $t = 1/4$ inch (minimum allowed by equation)

$$\sigma = \frac{Mc}{I} = \frac{M}{\pi t r^2} = \frac{400,000}{\pi \times 0.25 \times 48^2}$$
$$= 2660 \, \text{psi.}$$

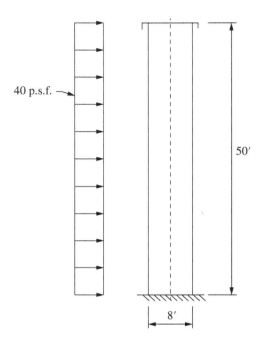

40 p.s.f. →

50′

8′

Figure 4.20   Low-carbon steel stack

From Eq. (4.73),

$$2\left(\frac{L}{r}\right) = 25.$$

Hence,

$$Y = 1.0$$

$$\left(\frac{t}{r}\right) = 0.0052.$$

Also,

$$X = [1,000,000 \times 0.25/48][2 - (0.667)(100 \times 0.25/48)]$$

$$X = 8600\,\text{psi.}$$

Allowable stress $= XY = 8600 \times 1.0 = 8600$ psi > 2660 psi.

Example 4.8
The flat bottom gasoline tank shown in Figure 4.21 is subjected to a snow load of 30 psf. Calculate the actual stress in the cylinder due to snow load and compare it to the allowable stress.

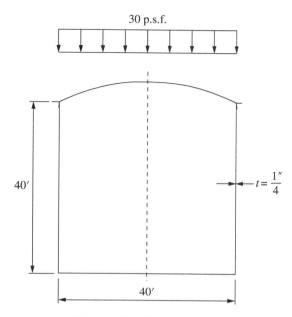

Figure 4.21   Flat bottom tank

Solution
Actual stress at the bottom of cylindrical shell

$$\sigma = \frac{p(\pi r^2)}{2\pi rt}$$
$$= \frac{(30/144)(240)}{2(0.25)} = 100\,\text{psi}.$$

From Eq. (4.74) with $t/r = 0.00104$,

$$\sigma = 1,800,000(0.00104) = 1870\,\text{psi} > 100\,\text{psi}.$$

Another methodology for calculating allowable compressive stress in a cylinder was developed by Miller (1999) and is used in Section VIII-2 of the ASME code. Miller combined experimental work he performed on flat bottom tanks with theoretical work and developed the following procedure for determining the allowable compressive stress in cylinders:

1. For $\lambda_c \leq 0.15$, calculate F_{xa1} and F_{xa2} as follows:
   For F_{xa1},

$$F_{xa1} = \frac{S_y}{\text{F.S.}} \quad \text{for} \quad \frac{D_o}{t} \leq 135$$

$$F_{xa1} = \frac{466 S_y}{[\text{F.S.}(331 + D_o/t)]} \quad \text{for} \quad 135 < \frac{D_o}{t} < 600$$

$$F_{xa1} = \frac{0.5S_y}{\text{F.S.}} \quad \text{for} \quad 600 \le \frac{D_o}{t} \le 2000$$

For F_{xa2},

$$\bar{c} = 2.64 \quad \text{for} \quad M_x \le 1.5$$

$$\bar{c} = \frac{3.13}{M_x^{0.42}} \quad \text{for} \quad 1.5 < M_x < 15$$

$$\bar{c} = 1.0 \quad \text{for} \quad M_x \ge 15$$

$$C_x = \min[0.9, 409\bar{c}/(389 + D_o/t)] \quad \text{for} \quad \frac{D_o}{t} < 1247$$

$$C_x = 0.25\bar{c} \quad \text{for} \quad 1247 \le \frac{D_o}{t} \le 2000$$

$$F_{xe} = \frac{C_x E_y t}{D_o}$$

$$F_{xa2} = \frac{F_{xe}}{\text{F.S.}}$$

The allowable compressive stress is given by

$$F_{xa} = \min[F_{xa1}, F_{xa2}] \tag{4.76}$$

2. For $\lambda_c > 0.15$ and $K_u L_u / r_g < 200$ (column buckling), calculate F_{ca} as follows:

$$F_{ca} = F_a[1 - 0.74(\lambda_c - 0.15)]^{0.3} \quad \text{for} \quad 0.15 < \lambda_c < 1.147 \tag{4.77a}$$

$$F_{ca} = 0.88 F_{xa}/\lambda_c^2 \quad \text{for} \quad \lambda_c \ge 1.147 \tag{4.77b}$$

where

$$\text{F.S.} = 2.0 \quad \text{for} \quad F_{1c} \le 0.55 S_y$$

$$= 2.407 - 0.741 \left(\frac{F_{1c}}{F_y}\right) \quad \text{for} \quad 0.55 S_y < F_{1c} < S_y$$

$$= 1.667 \quad \text{for} \quad F_{1c} = S_y$$

D_o = outside diameter of shell
F_{1c} = predicted buckling stress with F.S. = 1.0
K_u = effective length of column
L_u = length of column or shell subjected to axial compression

$$\lambda_c = \frac{K_u L_u}{\pi r_g} \left(F_{xa} \text{F.S.}/E_y\right)^{0.5}$$

$M_x = L/(R_o t)^{0.5}$

r_g = radius of gyration. For a thin shell, $r_g = R_o/1.41$

R_o = outside radius of shell

S_y = yield strength

t = thickness of shell

Example 4.9

Calculate the allowable axial compressive stress for a cylinder with $L = 368$ inch, $D_o = 48$ inch, $R_o = 24$ inch, $t = 0.375$ inch, $E = 30,000$ ksi, $S_y = 38$ ksi, $K_u = 2.1$, and $\mu = 0.3$ using Eqs. (4.76) and (4.77).

Solution

Assume factor of safety F.S. = 1.0.

Assume $\lambda_c \leq 0.15$.

$$\frac{D_o}{t} = \frac{48}{0.375} = 128.$$

$$F_{xa1} = \frac{38}{1} = 38 \text{ ksi.}$$

$$M_x = \frac{368}{[(24)(0.375)]^{0.5}} = 122.7.$$

$$\bar{c} = 1.0.$$

$$C_x = \min[0.9, (409)(1.0)/(389 + 128)] = 0.791.$$

$$F_{xe} = (0.791)(30,000)(0.375)/48 = 185 \text{ ksi.}$$

$$F_{xa2} = \frac{185}{1.0} = 185 \text{ ksi.}$$

From Eq. (4.76),

$$F_{xa} = \min[38, 185] = 38 \text{ ksi.}$$

Hence, $F_{1c} = 38.0$ ksi and the applicable factor of safety is 1.667.

Assume factor of safety F.S. = 1.667.

Assume $\lambda_c \leq 0.15$.

$$\frac{D_o}{t} = \frac{48}{0.375} = 128.$$

$$F_{xa1} = \frac{38}{1.667} = 22.8 \text{ ksi.}$$

$$M_x = \frac{368}{[(24)(0.375)]^{0.5}} = 122.7.$$

$$\bar{c} = 1.0.$$

$$C_x = \min[0.9, (409)(1.0)/(389 + 128)] = 0.791.$$

$$F_{xe} = (0.791)(30,000)(0.375)/48 = 185 \text{ ksi.}$$

$$F_{xa2} = \frac{185}{1.667} = 111 \text{ ksi.}$$

From Eq. (4.76),

$$F_{xa} = \min [22.8, 111] = 22.8 \text{ ksi.}$$

Check λ_c.

$$r_g = \frac{24}{1.4.1} = 16.97.$$

$$\lambda_c = \frac{(2.1)(368)}{\pi(16.97)} [(22.8)(1.667)/30,000]^{0.5} = 0.52.$$

Since this λ_c is greater than the assumed λ_c of 0.15, Eq. (4.77) is used.

$$K_u L_u / r_g = (2.1)(368)/16.97 = 45.5.$$

$$F_{ca} = 22.8[1 - 0.74(0.52 - 0.15)]^{0.3} = 22.7 \text{ ksi.}$$

Hence, the allowable compressive stress = 22.7 ksi.

Problems

4.1 Calculate the required thickness of the inner cylindrical shell using Eq. (4.45). Let $p = 15$ psi. $L = 20$ ft, $r = 20$ inches, F.S. = 3.0, $E = 15,000$ ksi, and $\mu = 0.3$. Assume the ends to be fixed.

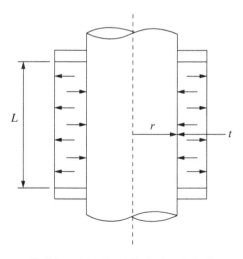

Problem 4.1 Partially jacketed shell

4.2 Find the required thickness of the inner cylinder shown due to a full vacuum condition, 15 psi, in the inner cylinder. Use Eq. (4.51). Let $E = 29,000$ ksi and $\mu = 0.30$. Use multiples of 1/16 inch in determining the cylinder thickness.

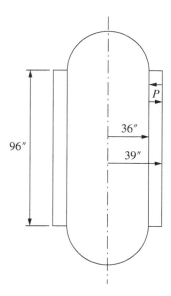

Problem 4.2   Jacketed pressure vessel

4.3 In Problem 4.2, find the required thickness in the outer cylinder due to a full vacuum condition in the annular jacket space. Use Eq. (4.51) even though the axial force on the outer cylinder is substantially less than that given by Eq. (4.51).

4.4 Solve Problem 4.3 using Eq. (4.45) as a more appropriate equation due to the small end load. What is the difference between this thickness and that obtained in Problem 4.3?

4.5 What is the required thickness of the support cylinder shown using Eq. (4.57)? The weight of the spherical tank is 100,000 lbs. Let $E = 30,000$ ksi, $\mu = 0.30$, factor of safety (F.S.) = 5.0. Use increments of 1/16 inch to determine thickness.

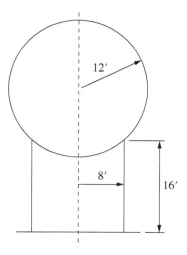

Problem 4.5   Support cylinder for a spherical fuel vessel

4.6 An oil storage standpipe has a thickness of 1/4″. Is the thickness adequate for axial compression due to deadweight? Let $E = 30{,}000$ ksi, $\mu = 0.30$, factor of safety (F.S.) = 5.0, and weight of steel = 490 pcf. Use Eq. (4.57).

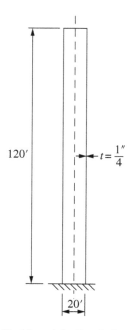

Problem 4.6   Stand pipe

4.7 Find the required thickness of the cylinder shown due to full vacuum condition. Use Eq. (4.62) or (4.63).

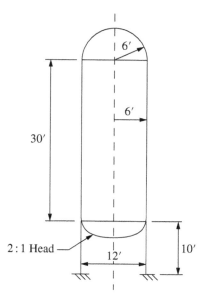

Problem 4.7   Process reactor

4.8 Solve Problem 4.7 using Eq. (4.65).

4.9 What is the required thickness of the supporting cylinder shown in Problem 4.7? The weight of the contents is 100,000 lbs.

4.10 What is the required thickness of the steel stack due to an effective wind load of 50 psf? Add the weight of the steel in the calculations. The thickness of the top 25 ft may be made different than the thickness of the bottom 25 ft. Weight of steel is 490 pcf. Use Eq. (4.73).

4.11 Calculate the required thickness of the cylindrical shell of the flat bottom tank shown in Figure 4.21 due to snow load of 30 psf plus the deadweight of the roof and cylinder. Use Eq. (4.74). Radius of the roof is 40 ft and the thickness is 0.375 inch.

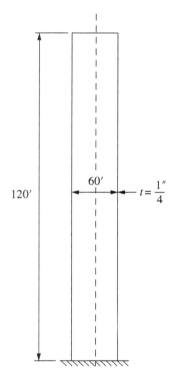

Problem 4.11 Steel stack

5

Stress in Shells of Revolution Due to Axisymmetric Loads

5.1 Elastic Stress in Thick-Wall Spherical Sections Due to Pressure

The required thickness of a thin spherical shell based on membrane theory is obtained from Eq. (1.13) as

$$t = \frac{PR}{2S} \tag{5.1}$$

where
P = pressure
R = inside radius
S = allowable stress
t = thickness

Equation (5.1) underestimates the actual stress in a heavy wall spherical section with a ratio of $R_i/t < 10$ due to the nonlinear stress distribution that develops across the wall of a thick spherical section due to internal pressure. In this chapter the equations for a heavy wall spherical section are presented for elastic, plastic, and creep conditions. The equations for the elastic analysis are derived similar to those for cylindrical shell in Chapter 3 by taking an infinitesimal cross section and summing forces (Faupel and Fisher 1981). The resulting stress equations are

$$\sigma_r = \frac{p_i\, R_i^3 - p_o\, R_o^3 - (p_i - p_o)\left(R_i^3\, R_o^3/r^3\right)}{R_o^3 - R_i^3} \tag{5.2}$$

Stress in ASME Pressure Vessels, Boilers, and Nuclear Components, First Edition. Maan H. Jawad.
© 2018, The American Society of Mechanical Engineers (ASME), 2 Park Avenue,
New York, NY, 10016, USA (www.asme.org). Published 2018 by John Wiley & Sons, Inc.

$$\sigma_\theta = \frac{p_i R_i^3 - p_o R_o^3 + (p_i - p_o)\left(R_i^3 R_o^3 / 2r^3\right)}{R_o^3 - R_i^3} \tag{5.3}$$

where

p_i = inside pressure
p_o = outside pressure
R_i = inside radius
R_o = outside radius
r = radius at any location
σ_r = radial stress
σ_θ = circumferential stress

The deflection, w, is expressed as

$$w = K_1 r + \frac{K_2}{r^2} \tag{5.4}$$

where

$$K_1 = \frac{1 - 2\mu}{E\left(R_o^3 - R_i^3\right)} \left(p_i R_i^3 - p_o R_o^3\right)$$

$$K_2 = \frac{1 + \mu}{2E\left(R_o^3 - R_i^3\right)} (p_i - p_o)\left(R_i^3 R_o^3\right)$$

and E = modulus of elasticity and μ = Poisson's ratio

The maximum stress due to internal pressure is obtained from Eq. (5.3) and occurs at the inside surface

$$\sigma_\theta = \frac{p_i\left(R_i^3 + R_o^3 / 2\right)}{R_o^3 - R_i^3}. \tag{5.5}$$

This equation, referred to as Lame's equation, can be written in terms of thickness, t, as

$$t = R_i \left\{ \left[\frac{2(S + p_i)}{(2S - p_i)}\right]^{1/3} - 1 \right\} \tag{5.6}$$

where S is the allowable stress and t is the thickness. This closed-form equation is convenient for calculating the required thickness of a spherical section once the inside radius, internal pressure, and allowable stress are known.

5.2   Spherical Shells in the ASME Code

5.2.1   Section I: Power Boilers

The design equation in Section I for spherical shells subjected to internal pressure is expressed as

$$t = \frac{PL}{2Sw - 0.2P} \tag{5.7}$$

where
L = inside radius
P = internal pressure
S = allowable stress in tension
t = thickness
w = weld joint strength reduction factor

Equation (5.7) is the same as Eqs. (1.13) and (1.14) with a $0.2P$ modifier in the denominator. The $0.2P$ modifier is an adjustment to account for the increase in stress level in thick heads to simulate that obtained from Eq. (5.5). The stress obtained from Eq. (5.7) is within 5% of that obtained from Eq. (1.14) for $L/t = 4.0$. The difference increases substantially for $L/t < 4.0$ with the results obtained from Eq. (1.14) being less conservative.

The rules in Section I of the ASME Code state that Eq. (5.7) cannot be used when $t \geq 0.356R$. In this case Lame's Eq. (5.5) is applicable.

5.2.2 Section III: Nuclear Components

5.2.2.1 Section NB

Section NB lists the following equation for calculating the required thickness in a spherical part:

$$t = \frac{PR}{2S - P} \tag{5.8}$$

where
P = internal pressure
R = inside radius
S = allowable stress in tension
t = thickness

This equation is obtained from the principal stress theory that states the maximum stress is equal to the largest difference in the three principal stresses: longitudinal, circumferential, and radial. The membrane stresses in the three principal axes for a thin head are

$$S_c = \frac{PR}{2t}, \quad S_L = \frac{PR}{2t}, \quad \text{and} \quad S_r = \frac{-P}{2}.$$

and the maximum principal stress is $S_m = (PR/2t) - (-P/2)$. This expression can be rearranged to give Eq. (5.8).

5.2.2.2 Section NC

The design equation in Section NC is given by

$$t = \frac{PR}{2S - 0.5P} \tag{5.9}$$

One other equation for calculating required thickness is given in Section NC when $P > 0.4S$ and is based on plastic analysis. This equation is discussed later in this chapter.

5.2.2.3 Section ND

The design equation in Section ND is given by

$$t = \frac{PR}{2SE - 0.2P} \tag{5.10}$$

where, E = weld joint efficiency.

Another equation for calculating required thickness is given in Section ND when $t > 0.356R$ or when $P > 0.665S$ and is based on Lame's Eq. (5.5) for thick heads.

5.2.2.4 Section NE

The design equations in Section NE are the same as those in Section ND.

5.2.3 Section VIII: Pressure Vessels

5.2.3.1 Division 1

The basic design equation given in VIII-1 is the same as that given in Section ND, Eq. (5.10). Another equation for calculating the required thickness is given in Section VIII-1 when $t > 0.356R$ or when $P > 0.665S$ and is based on plastic analysis as discussed later in this chapter.

5.2.3.2 Division 2

The design equation given in VIII-2 is based on plastic analysis. This equation is discussed later in this chapter.

Example 5.1

A spherical head with an inside radius of 20 inches and allowable stress of 20,000 psi is subjected to an internal pressure of 3000 psi. Find the required thickness based on:

a. Thin shell Eq. (5.1).
b. Lame's Eq. (5.6).
c. Equation (5.7) from Section I. Let $w = 1.0$.
d. Equation (5.8) from Section III-NB.
e. Equation (5.9) from Section III-NC.
f. Equation (5.10) from Sections III-ND and Section VIII-1. Let $E = 1.0$.

Solution

a. $t = PR/2S = (3000)(20)/40,000 = 1.50$ inch.
b. $t = (20)\{[2(20,000 + 3000)/(2(20,000) - 3000)]^{1/3} - 1\} = 1.51$ inch.

c. $t = \dfrac{(3000)(20)}{2(20,000)(1.0) - 0.2(3000)} = 1.523$ inch.

d. $t = \dfrac{(3000)(20)}{2(20,000) - (3000)} = 1.622$ inch.

e. $t = \dfrac{(3000)(20)}{2(20,000) - 0.5(3000)} = 1.558$ inch.

f. $t = \dfrac{(3000)(20)}{2(20,000)(1.0) - 0.2(3000)} = 1.523$ inch.

5.3   Stress in Ellipsoidal Shells Due to Pressure Using Elastic Analysis

The maximum stress in a thin ellipsoidal head based on membrane theory occurs at the apex of the head. The required thickness is obtained from Eq. (1.20) as

$$t = \frac{Pa^2}{2bS} \tag{5.11}$$

where
a = major radius
b = minor radius
P = pressure
S = allowable stress
t = thickness

For a $2:1$ head with $a = 2b$ and $a = D/2$, Eq. (5.11) becomes

$$t = PD/2S. \tag{5.12}$$

For a flanged and dished head (6% knuckle) with $a = 2.95b$, and $a = D/2$, Eq. (5.11) becomes

$$t = 0.74PD/S. \tag{5.13}$$

Equations (5.12) and (5.13) are applicable at the apex of the head. Experience has shown that the thickness obtained from these two equations may not be adequate at the knuckle region that is in compression due to internal pressure as shown in Figure 1.14. Accordingly, buckling may need to be factored into the design. A common procedure is to use Figure 5.1 to calculate the required thickness of shallow heads that take into consideration buckling of the knuckle region. The procedure consists of calculating the quantity P/S and then using Figure 5.1 to obtain a quantity t/L from which the thickness is determined. In lieu of using the figure, the following equation may be utilized:

$$t = Le^A \tag{5.14}$$

where

$$A = \left(a_1 + a_2x + a_3x^2\right) + \left(b_1 + b_2x + b_3x^2\right)y + \left(c_1 + c_2x + c_3x^2\right)y^2$$

D = base diameter of head
L = spherical radius
r = knuckle radius
S = allowable stress
t = thickness
$x = r/D$
$y = \ln(P/S)$

and

$$a_1 = -1.2617702 \quad a_2 = -4.5524592 \quad a_3 = 28.933179$$
$$b_1 = 0.66298796 \quad b_2 = -2.2470836 \quad b_3 = 15.682985$$
$$c_1 = 0.26878909E-04 \quad c_2 = -0.42262179 \quad c_3 = 1.8878333$$

Section VIII-2 uses equations for design rules, while Section III uses a chart. This chart is shown in Figure 5.1. The design is based on approximating the geometry of a head with a spherical radius, L, and a knuckle radius, r, as defined in Figure 5.1. The required thickness of a specific head is determined from Figure 5.1 by knowing the values of L, r, base diameter of the head, applied pressure, and allowable membrane stress.

Example 5.2
Some of the commonly used boiler heads in the United States are ellipsoidal in shape and have an a/b ratio of about 2.95. This corresponds to an approximate value, Figure 1.13, of $L = D$ and $r = 0.06D$. Determine the required thickness of this head if $D = 14$ ft, $p = 75$ psi, and $\sigma = 20,000$ psi. Calculate the required thickness from Eq. (1.18) of and also from Figure 5.1.

Solution
From Eq. (1.18), the maximum force is at $\phi = 0°$. Thus,

$$t = \frac{pa^2}{2b\sigma}$$

$$= \frac{75 \times (168/2)^2}{2 \times (84/2.95) \times 20,000}$$

$$= 0.46 \,\text{inch.}$$

From Figure 5.1 with $p/\sigma = 0.0038$ and $r/D = 0.06$, we get $t/L = 0.005$, or $t = 0.005 \times 168 = 0.84$ inch (controls).

Notice that the thickness obtained from Figure 5.1 is almost double than that determined from theoretical membrane equations that do not take into consideration any instability due to internal pressure.

5.4 Ellipsoidal (Dished) Heads in the ASME Code

5.4.1 Section I: Power Boilers

The design equation given in Section I for dished heads is

$$t = \frac{5PL}{4.8Sw} \tag{5.15}$$

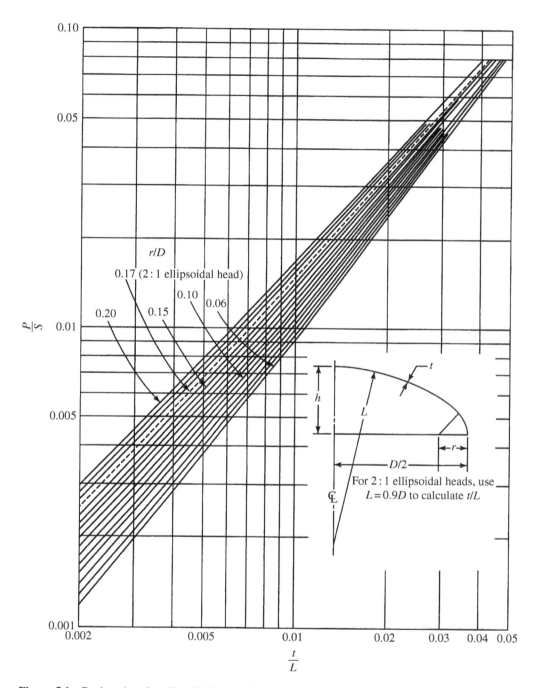

Figure 5.1 Design chart for ellipsoidal heads. Source: ASME. Reproduced with permission of ASME

where
L = inside radius of dished head
P = internal pressure
S = allowable stress in tension
t = thickness
w = weld joint strength reduction factor

For a $2:1$ ellipsoidal head, Figure 1.12, where $L = 0.9D$, this equation becomes

$$t = \frac{0.94PD}{Sw} \qquad (5.16)$$

where, D = base diameter of head
  The design equation for an F&D head, Figure 1.13, where $L = D$ is

$$t = \frac{1.04PD}{Sw}. \qquad (5.17)$$

5.4.2   Section III: Nuclear Components

5.4.2.1   Section NC

The design equation for ellipsoidal and F&D heads in Section NC is based on Figure 5.1.

5.4.2.2   Section ND

The design equation for $2:1$ ellipsoidal head in Section ND is given by

$$t = \frac{PD}{2SE - 0.2P} \qquad (5.18)$$

where
E = weld joint efficiency
P = internal pressure
D = inside diameter
S = allowable stress in tension
t = thickness

This equation is essentially the same as the theoretical Eq. (5.12) with a modification to account for thick shell theory.
  The design equation for F&D heads in Section ND is given by

$$t = \frac{0.885PL}{SE - 0.1P}. \qquad (5.19)$$

This equation results in thicknesses that are 20% higher than those calculated from Eq. (5.13).

Section ND also gives modifiers to Eqs. (5.18) and (5.19) for shapes other than 2 : 1 and F&D geometries.

5.4.2.3   Section NE

The design equations in Section NE are the same as those in Section ND.

5.4.3   Section VIII: Pressure Vessels

5.4.3.1   Division 1

The basic design equations for ellipsoidal and torispherical heads given in Section VIII-1 are the same as those given in Section ND. For thin heads, the design is based on Eq. (5.14).

5.4.3.2   Division 2

The design equations take into consideration the stress at the apex per Eq. (5.11) plus the stress at the knuckle per Eq. (5.14).

Example 5.3

An F&D head with an inside diameter of 14 ft and allowable stress of 20,000 psi is subjected to an internal pressure of 75 psi. Find the required thickness based on

a. Thin shell Eq. (5.13).
b. Equation (5.14).
c. Equation (5.17) from Section I. Let $w = 1.0$.
d. Equation (5.19) from Sections III-ND and VIII-1. Let $E = 1.0$.

Solution
a. The thickness from Eq. (5.13) is

$$t = \frac{(0.74)(75)(168)}{20\ 000} = 0.466 \text{ inch.}$$

b. The thickness from Eq. (5.14) is calculated as follows:

$$L = 168 \text{ inch} \qquad r = 10.08 \text{ inch}$$
$$x = 0.06 \qquad\qquad y = -5.586 \qquad A = -5.27478$$
$$t = (168)(e^{-5.27478}) = 0.86 \text{ inch}$$

Hence, compressive forces in the knuckle rather than tensile forces near the crown control the thickness of the head.

c. $t = \dfrac{1.04(75)(168)}{(20,000)(1.0)} = 0.655 \text{ inch.}$

d. $t = \dfrac{0.885(75)(168)}{(20,000)(1.0) - 0.1(75)} = 0.558 \text{ inch.}$

5.5 Stress in Thick-Wall Spherical Sections Due to Pressure Using Plastic Analysis

The plastic analysis of spherical shells follows along the same lines as those in Section 3.4 for cylindrical shells. The resulting design equation for a fully plastic condition is similar to Eq. (3.39) and is of the form

$$t = R_i \left(e^{0.5P/S} - 1 \right). \tag{5.20a}$$

The circumferential stress for a fully plastic condition is given by the equation

$$S_\theta = S \left[1 + \ln \left(\frac{r}{R_o} \right) \right] \tag{5.20b}$$

where, S_θ = actual circumferential stress and S = allowable stress.

Example 5.4
Find the thickness of a spherical head welded to the shell in Example 3.3 using plastic analysis.

Solution
From Eq. (5.20)

$$t = 12 \left(e^{0.5(9\,615)/25,000} - 1 \right)$$
$$t = 2.54 \text{ inch}$$

5.6 Stress in Thick-Wall Spherical Sections Due to Pressure Using Creep Analysis

The expression for stress due to internal pressure in a thick cylinder operating in the creep range (Jones 2009) is based on Norton's equation and is similar to that in Section 3.5 for cylindrical shell. It is given by

$$\sigma_\theta = p_i \frac{[(3-2n)/2n](R_o/r)^{3/n} + 1}{(R_o/R_i)^{3/n} - 1}. \tag{5.21}$$

Example 5.5
Find the maximum stress of the head in Example 5.4 using creep theory. Let $n = 5.1$.

Solution
From Eq. (5.21) it is seen the maximum stress occurs at $r = R_o$. Hence,

$$\sigma_\theta = (9615) \frac{[(3-2(5.1))/2(5.1)](14.54/14.54)^{3/5.1} + 1}{(14.54/12)^{3/5.1} - 1}.$$

$$\sigma_\theta = 23,650 \text{ psi}.$$

5.7   Bending of Shells of Revolution Due to Axisymmetric Loads

In Chapter 1 the basic equations of equilibrium for the membrane forces in shells of revolution due to axisymmetric loads were developed. Referring to Figure 1.4, it was shown that the two governing equations of equilibrium are given by Eqs. (1.7) and (1.9) as

$$\frac{d}{d\phi}(rN_\phi) - r_1 N_\theta \cos\phi + p_\phi r r_1 = 0 \tag{5.22}$$

and

$$\frac{N_\phi}{r_1} + \frac{N_\theta}{r_2} = p_r. \tag{5.23}$$

In many applications such as at a junction of a spherical-to-cylindrical shell subjected to axisymmetric loads, bending moments and shear forces are developed at the junction in order to maintain equilibrium and compatibility between the two shells. The effect of these additional moments and shears, Figure 5.2, on a shell of revolution (Gibson 1965) is the subject of this chapter. The axisymmetric moments and shears at the two circumferential edges of the infinitesimal element are shown in Figure 5.2. Circumferential bending moments, which are constant in the θ-direction for any given angle ϕ, are applied at the meridional edges of the element. The shearing forces at the meridional edges must be zero in order for the deflections, which must be

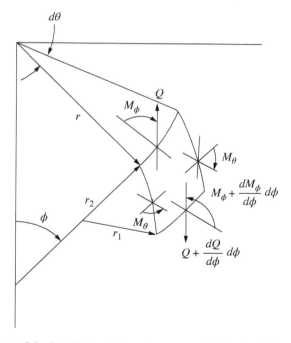

Figure 5.2   Bending and shear forces on an infinitesimal element

symmetric in the θ-direction because the loads are axisymmetric, to be constant in the θ-direction for any given angle ϕ.

Summation of forces in Figure 5.2 parallel to the tangent at the meridian results in

$$Q(rd\theta)\sin d\phi/2 + \left(Q + \frac{dQ}{d\phi}d\phi\right)\left(r + \frac{dr}{d\phi}d\phi\right)d\theta\sin d\phi/2 = 0.$$

Simplifying this equation gives

$$Qr\, d\phi\, d\theta = 0.$$

Adding this expression to Eq. (5.22) gives

$$\frac{d}{d\phi}(rN_\phi) - r_1 N_\theta \cos\phi - Qr + p_\phi rr_1 = 0. \tag{5.24}$$

Summation of forces perpendicular to the middle surface in Figure 5.2 gives

$$-Q(rd\theta)\cos d\phi/2 + \left(Q + \frac{dQ}{d\phi}d\phi\right)\left(r + \frac{dr}{d\phi}d\phi\right)d\theta\cos d\phi/2 = 0.$$

Simplifying this expression and adding it to Eq. (5.23) gives

$$N_\phi r + N_\theta r_1 \sin\phi + \frac{d(Qr)}{d\phi} - p_r rr_1 = 0. \tag{5.25}$$

Summation of moments in the direction of a parallel circle gives

$$\frac{d}{d\phi}(M_\phi r) - M_\theta r_1 \cos\phi - Qr_1 r_2 \sin\phi = 0. \tag{5.26}$$

The second term involving M_θ in this equation is obtained from Figure 5.3.

Equations (5.24) through (5.26) contain the five unknowns N_θ, N_ϕ, Q, M_θ, and M_ϕ. Accordingly, additional equations are needed and are obtained from the relationship between deflections and strains. The expressions for the meridional and circumferential strains were obtained in Chapter 2 as Eqs. (2.2) and (2.3) and are given by

$$\varepsilon_\phi = \left(\frac{(dv/d\phi) - w}{r_1}\right) \tag{5.27}$$

$$\varepsilon_\theta = \frac{v\cot\phi - w}{r_2}. \tag{5.28}$$

The expressions for N_θ and N_ϕ are obtained from

$$N_\phi = \frac{Et}{1-\mu^2}(\varepsilon_\phi + \mu\varepsilon_\theta) \tag{5.29}$$

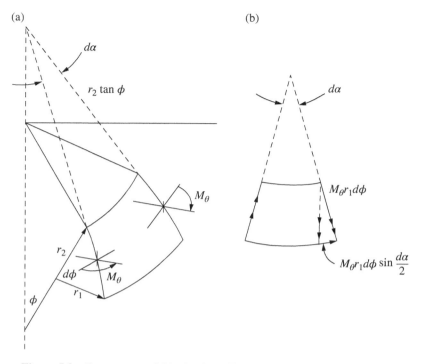

Figure 5.3 Components of M_θ: (a) three-dimensional view and (b) side view

$$N_\theta = \frac{Et}{1-\mu^2}\left(\varepsilon_\theta + \mu\varepsilon_\phi\right). \tag{5.30}$$

Substituting Eqs. (5.27) and (5.28) into Eqs. (5.29) and (5.30) gives

$$N_\phi = \frac{Et}{1-\mu^2}\left[\frac{1}{r_1}\left(\frac{dv}{d\phi}-w\right) + \frac{\mu}{r_2}(v\cot\phi - w)\right] \tag{5.31}$$

$$N_\theta = \frac{Et}{1-\mu^2}\left[\frac{1}{r_2}(v\cot\phi - w) + \frac{\mu}{r_1}\left(\frac{dv}{d\phi}-w\right)\right]. \tag{5.32}$$

The expression for change of curvature in the ϕ-direction is obtained from Figure 5.4. The rotation α of point A in Figure 5.4 is the summation of rotation α_1 due to deflection v and rotation α_2 due to deflection w, Figure 5.5. From Figure 5.5a,

$$\alpha_1 = \frac{v}{r_1}$$

and from Figure 5.5b,

$$\alpha_2 = \frac{dw}{r_1 d\phi}.$$

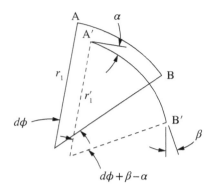

Figure 5.4 Deflection of a shell segment

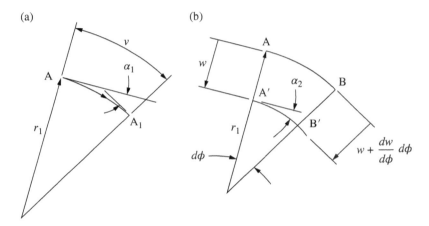

Figure 5.5 Rotation of a shell segment: (a) rotation of element and (b) deflection of element

Hence,

$$\alpha = \alpha_1 + \alpha_2 = \left(\frac{v + (dw/d\phi)}{r_1} \right). \tag{5.33}$$

Similarly the rotation β of point B is the summation of rotations due to deflection v and w. The rotation due to deflection v is expressed as

$$\left(\frac{v + (dv/d\phi)d\phi}{r_1} \right)$$

and the rotation due to w is

$$\frac{dw}{r_1 d\phi} + \frac{d}{d\phi}\left(\frac{dw}{r_1 d\phi} \right)d\phi.$$

Hence,

$$\beta = \left(\frac{v + (dv/d\phi)d\phi}{r_1}\right) + \frac{dw}{r_1 \, d\phi} + \frac{d}{d\phi}\left(\frac{dw}{r_1 \, d\phi}\right)d\phi. \tag{5.34}$$

Due to rotation, the middle surface does not change in length. Thus, from Figure 5.4

$$AB = A'B'$$

$$r_1 \, d\phi = r_1 (dQ + \beta - \alpha)$$

or

$$\frac{1}{r_1'} = \frac{d\phi + \beta - \alpha}{r_1 \, d\phi}.$$

Change in curvature

$$\chi_\phi = \frac{1}{r_1'} - \frac{1}{r_1}$$

$$= \frac{(\beta - \alpha)}{r_1 \, d\phi} \tag{5.35}$$

$$\chi_\phi = \frac{d}{r_1 \, d\phi}\left(\frac{v}{r_1} + \frac{dw}{r_1 \, d\phi}\right).$$

The change of curvature in the θ-direction is obtained from Figure 5.6, which shows the rotation of side AB due to the deformation of element $ABCD$. The original length AB is given by

$$AB = r_2 \sin \phi \, d\theta.$$

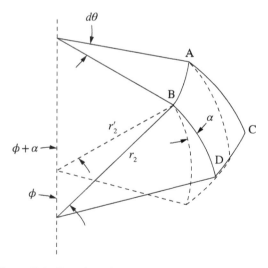

Figure 5.6   Rotation of a shell element in the θ-direction

After rotation, AB is expressed by

$$AB = r_2' \sin(\phi + \alpha)\, d\theta.$$

Equating these two expressions and assuming small angle rotation yields

$$\frac{1}{r_2'} = \frac{1}{r_2}(1 + \alpha \cot \phi)$$

and from

$$\chi_\theta = \frac{1}{r_2'} - \frac{1}{r_2}$$

we get

$$\chi_\theta = \frac{\alpha \cot \phi}{r_2}.$$

Substituting Eq. (5.33) into this expression gives

$$\chi_\theta = \frac{1}{r_2}\left(\frac{v}{r_1} + \frac{dw}{r_1\, d\phi}\right)\cot \phi. \tag{5.36}$$

The relationship between moment and rotation is obtained from strength of materials and given by Eqs. (7.6), (7.7), and (7.17) as

$$M_\phi = D(\chi_\phi + \mu \chi_\theta) \tag{5.37}$$

$$M_\theta = D(\chi_\theta + \mu \chi_\phi). \tag{5.38}$$

Substituting Eqs. (5.35) and (5.36) into these two expressions gives

$$M_\phi = D\left(\frac{1}{r_1}\frac{d\alpha}{d\phi} + \frac{\mu}{r_2}\alpha \cot \phi\right) \tag{5.39}$$

$$M_\theta = D\left(\frac{1}{r_2}\alpha \cot \phi + \frac{\mu}{r_1}\frac{d\alpha}{d\phi}\right). \tag{5.40}$$

The eight equations (5.24), (5.25), (5.26), (5.31), (5.32), (5.33), (5.39), and (5.40) contain eight unknowns. They are N_θ, N_ϕ, Q, M_ϕ, M_θ, v, w, and α. Solution of these equations is discussed next.

5.8   Spherical Shells

The forces and moments throughout a spherical shell due to edge shears and moments will be derived in this section. For spherical shells, Figure 5.2, the radii r_1 and r_2 are equal to R. Also, the pressures p_r and p_ϕ are set to zero for the case of applied edge loads and moments. The eight pertinent equations derived in Section 5.7 can now be reduced to two differential equations.

The first equation is obtained by substituting the moment Eqs. (5.39) and (5.40) to Eq. (5.26). This gives

$$\frac{d^2\alpha}{d\phi^2} + \frac{d\alpha}{d\phi}\cot\phi - \alpha\left(\cot^2\phi + \mu\right) = -\frac{QR^2}{D}. \tag{5.41}$$

The second differential equation is more cumbersome to derive. We start by substituting Eq. (5.24) into Eq. (5.25) to delete N_θ. Then integrating the resultant equation with respect to ϕ gives

$$N_\phi = -Q\cot\phi. \tag{5.42}$$

Substituting this expression into Eq. (5.25) yields

$$N_\theta = -\frac{dQ}{d\phi}. \tag{5.43}$$

From Eqs. (5.31) and (5.32), we get

$$\frac{dv}{d\phi} - w = \frac{R(N_\phi - \mu N_\theta)}{Et} \tag{5.44}$$

and

$$v\cot\phi - w = \frac{R(N_\theta - \mu N_\phi)}{Et}. \tag{5.45}$$

Combining Eqs. (5.44) and (5.45) results in

$$\frac{dv}{d\phi} - v\cot\phi = \frac{R(1+\mu)(N_\phi - N_\theta)}{Et}. \tag{5.46}$$

Differentiating Eq. (5.45) and combining it with Eq. (5.46) gives the expression

$$v + \frac{dw}{d\phi} = \frac{R}{Et}\left[(1+\mu)(N_\phi - N_\theta) - \frac{dN_\theta}{d\phi} - R\frac{dN_\phi}{d\phi}\right].$$

Substituting Eqs. (5.42) and (5.43) into this expression results in

$$\frac{d^2Q}{d\phi^2} + \frac{dQ}{d\phi}\cot\phi - Q\left(\cot^2\phi - \mu\right) = Et\alpha. \tag{5.47}$$

Equations (5.41) and (5.47) must be solved simultaneously to determine Q and α. The exact solution of these two equations is too cumbersome to use for most practical problems and is beyond the scope of this book. Timoshenko and Woinowsky-Krieger (1959) showed that a rigorous solution of Eqs. (5.41) and (5.47) results in expressions for α and Q that contain the terms $e^{\lambda\phi}$ and $e^{-\lambda\phi}$ where λ is a function of R/t. These terms have a large numerical value for thin shells with large R/t ratios. Substitution of these terms into Eqs. (5.41) and (5.47) for shells with large ϕ angles results in two equations with substantially larger numerical values for the higher derivatives $d^2Q/d\phi^2$ and $d^2\alpha/d\phi^2$ compared with the other terms in the equations. Hence, a reasonable approximation of Eqs. (5.41) and (5.47) is obtained by rewriting them as

$$\frac{d^2\alpha}{d\phi^2} = -\frac{QR^2}{D} \tag{5.48}$$

and

$$\frac{d^2Q}{d\phi^2} = Et\alpha. \tag{5.49}$$

Substituting Eq. (5.48) into Eq. (5.49) gives

$$\frac{d^4Q}{d\phi^4} + 4\lambda^4 Q = 0 \tag{5.50}$$

where

$$\left.\begin{aligned}\lambda^4 &= EtR^2/4D \\ &= 3(1-\mu^2)(R/t)^2.\end{aligned}\right] \tag{5.51}$$

The solution of Eq. (5.50) is similar to that obtained for cylindrical shells and is given by either

$$Q = e^{\lambda\phi}(C_1\sin\lambda\phi + C_2\cos\lambda\phi) + e^{-\lambda\phi}(C_3\sin\lambda\phi + C_4\cos\lambda\phi) \tag{5.52}$$

or

$$\begin{aligned}Q = K_1\sin\lambda\phi\sinh\lambda\phi + K_2\sin\lambda\phi\cosh\lambda\phi + K_3\cos\lambda\phi\sinh\lambda\phi \\ + K_4\cos\lambda\phi\cosh\lambda\phi.\end{aligned} \tag{5.53}$$

For continuous spherical shells without holes and subjected to edge forces, the constants C_3 and C_4 in Eq. (5.52) must be set to zero in order for Q to diminish as ϕ gets smaller. Hence, Eq. (5.52) becomes

$$Q = e^{\lambda\phi}(C_1\sin\lambda\phi + C_2\cos\lambda\phi). \tag{5.54}$$

This equation can be written in a more compact form by substituting for ϕ the quantity $(\phi_0 - \zeta)$ shown in Figure 5.7. The new equation with redefined constants is

$$Q = C_1 e^{-\lambda\zeta}\sin(\lambda\zeta - C_2). \tag{5.55}$$

After obtaining Q, other forces and moments can be determined. Thus, from Eq. (5.42), we obtain the longitudinal membrane force

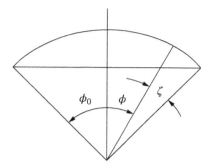

Figure 5.7 Designation of angles in a spherical shell

$$N_\phi = -C_1 e^{-\lambda\zeta} \sin(\lambda\zeta + C_2)\cot(\phi_o - \zeta). \tag{5.56}$$

The circumferential membrane force is determined from Eq. (5.43) as

$$N_\theta = -\sqrt{2}C_1 e^{-\lambda\zeta} \sin(\lambda\zeta + C_2 - \pi/4). \tag{5.57}$$

The rotation is determined from Eq. (5.49),

$$\alpha = -\frac{2\lambda^2}{Et} C_1 e^{-\lambda\zeta} \cos(\lambda\zeta + C_2). \tag{5.58}$$

The moments are obtained from Eqs. (5.39) and (5.40) using only higher order derivatives

$$\left.\begin{aligned} M_\phi &= -\frac{D\,d\alpha}{R\,d\phi} \\ &= \frac{R}{\sqrt{2}\lambda} C_1 e^{-\lambda\zeta} \sin(\lambda\zeta + C_2 + \pi/4) \end{aligned}\right] \tag{5.59}$$

$$M_\theta = \mu M_\phi. \tag{5.60}$$

The horizontal displacement is obtained from Figure 5.8 as

$$w_h = v\cos\phi - w\sin\phi$$
$$= (v\cot\phi - w)\sin\phi \tag{5.61}$$

and from Eq. (5.45)

$$w_h = \frac{R}{Et}(N_\theta - \mu N_\phi)\sin\phi \tag{5.62}$$

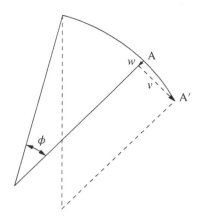

Figure 5.8 Displacement of spherical shell

$$= -\frac{\sqrt{2}\lambda R}{Et} C_1 e^{-\lambda\zeta}\sin(\phi_o - \zeta)\sin(\lambda\zeta + C_2 - \pi/4). \tag{5.63}$$

Table 5.1 gives various design values for spherical shells due to applied edge loads.

Example 5.6
Find the location and maximum value of moment M_ϕ due to the horizontal force shown in Figure 5.9.

Solution
The first boundary condition at the edge is

$$Q = H_o \sin\phi_o \quad \text{at} \quad \zeta = 0$$

and from Eq. (5.55)

$$C_1 \sin(-C_2) = H_o \sin\phi_o. \tag{1}$$

From Eqs. (5.49) and (5.59)

$$M = \frac{-D}{R}\frac{1}{Et}\frac{d^3 Q}{d\zeta^3}.$$

Substituting the third derivative of Eq. (5.55) into this expression gives

$$M_\phi = \frac{-2DC_1\lambda^3 e^{-\lambda\zeta}}{REt}[(\cos C_2)(\cos\lambda\zeta + \sin\lambda\zeta) + (\sin C_2)(\sin\lambda\zeta - \cos\lambda\zeta)].$$

The second boundary condition at the edge is

$$M_o = 0$$

Table 5.1 Horizontal edge load on a spherical shell

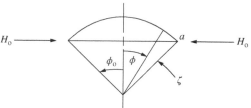

Edge functions $\zeta = 0$ and $\phi = \phi_o$

$$\alpha \quad -\frac{2H_o\lambda^2}{Et}\sin\phi_o$$

$$w_h \quad \frac{H_oR}{Et}\sin\phi_o(2\lambda\sin\phi_o - \mu\cos\phi_o)$$

General functions

$$Q \quad \sqrt{2}H_oe^{-\lambda\zeta}\sin\phi_o\cos\left(\lambda\zeta + \frac{\pi}{4}\right)$$

$$N_\theta \quad -2H_o\lambda e^{-\lambda\zeta}\sin\phi_o\cos\lambda\zeta$$

$$M_\phi \quad -\frac{H_oR}{\lambda}e^{-\lambda\zeta}\sin\phi_o\sin\lambda\zeta$$

$$\alpha \quad -\frac{H_o}{Et}\left[2\sqrt{2}\lambda^2e^{-\lambda\zeta}\sin\phi_o\sin\left(\lambda\zeta + \frac{\pi}{4}\right)\right]$$

$$w_h \quad \frac{H_oR}{Et}e^{-\lambda\zeta}\sin\phi_o\left[2\lambda\sin\phi\cos\lambda\zeta - \sqrt{2}\mu\cos\phi\cos\left(\lambda\zeta + \frac{\pi}{4}\right)\right]$$

Bending edge moment on a spherical shell

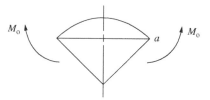

Edge functions $\zeta = 0$ and $\phi = \phi_o$

$$\alpha \quad \frac{4M_o\lambda^3}{EtR}$$

$$w_h \quad -\frac{2M_o\lambda^2}{Et}\sin\phi_o$$

General functions

$$Q \quad -\frac{2M_o\lambda}{R}e^{-\lambda\zeta}\sin\lambda\zeta$$

$$N_\theta \quad 2\sqrt{2}\frac{M_o\lambda^2}{R}e^{-\lambda\zeta}\cos\left(\lambda\zeta + \frac{\pi}{4}\right)$$

$$M_\phi \quad \sqrt{2}M_oe^{-\lambda\zeta}\sin\left(\lambda\zeta + \frac{\pi}{4}\right)$$

$$\alpha \quad \frac{4M_o\lambda^3}{EtR}e^{-\lambda\zeta}\cos\lambda\zeta$$

$$w_h \quad -\frac{2M_o\lambda}{Et}e^{-\lambda\zeta}\left[\sqrt{2}\lambda\sin\phi\cos\left(\lambda\zeta + \frac{\pi}{4}\right) + \mu\cos\phi\sin\lambda\zeta\right]$$

Notation: $M_\theta = \mu M_\phi$; $N_\phi = -Q\cot\phi$; w_h = horizontal component of deflection; α = rotation; $\zeta = \phi_o - \phi$; $\lambda = \sqrt[4]{(R/t)^2 3(1-\mu^2)}$; μ = Poisson's ratio. Inward deflection is positive. Positive rotation is in direction of positive moments. Tensile N_ϕ and N_θ are positive. Positive moments cause tension on the inside surface. Inward Q is positive.

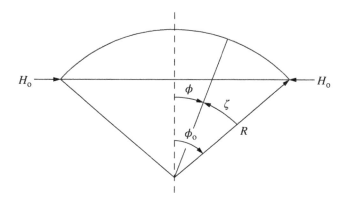

Figure 5.9 Horizontal edge load

or

$$\cos C_2(1) + \sin C_2(-1) = 0. \tag{2}$$

From Eq. (2) we get

$$C_2 = \frac{\pi}{4}$$

and from Eq. (1) we get

$$C_1 = \frac{2}{\sqrt{2}} H_o \sin \phi_o.$$

The expression for M_ϕ becomes

$$M_\phi = \frac{4DH_o\lambda^3 e^{-\lambda\zeta}}{REt} \sin\phi \sin\lambda\zeta.$$

Substituting

$$D = \frac{Et^3}{12(1-\mu^2)}$$

into the moment expression results in

$$M_\phi = \frac{H_oR}{\lambda} e^{-\lambda\zeta} \sin\phi_o \sin\lambda\zeta \tag{3}$$

which is the same as that given in Table 5.1.

The location of the maximum moment is obtained from

$$\frac{dM_\phi}{d\zeta} = 0$$

or

$$\zeta = \frac{\pi}{2\lambda}.$$

The maximum moment is obtained from Eq. (3) as

$$M_o = \frac{0.2079 H_o R}{\lambda} \sin\phi_o.$$

Example 5.7

Calculate the required thickness of the hemispherical and cylindrical shells in Figure 5.10a and determine the discontinuity stress at the junction. Let $p = 200$ psi. Allowable membrane stress is 18 ksi and $\mu = 0.3$.

Solution

The required thickness of the cylindrical shell is

$$t = \frac{pr}{\sigma} = \frac{200 \times 36}{18,000} = 0.40 \text{ inch.}$$

The required thickness of the hemispherical shell is

$$t = \frac{pR}{2\sigma} = \frac{200 \times 36}{2 \times 18,000} = 0.20 \text{ inch.}$$

The discontinuity forces are shown in Figure 5.10b. The first compatibility equation at point a is

deflection of cylinder due to $p + H_o + M_o$ = deflection of hemisphere due to $p + H_o + M_o$ (1)

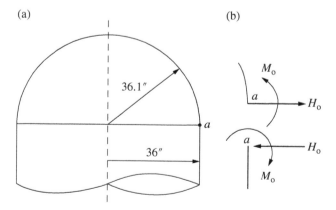

Figure 5.10 Hemispherical-to-cylindrical shell junction: (a) head-to-shell junction and (b) discontinuity forces

The expressions for the deflection of the cylinder are obtained from Eq. (3.95) and Table 3.1. For the hemisphere, the expressions are obtained from Eq. (2.11) and Table 5.1. Let outward deflection and clockwise rotation at point a be positive. Hence,

$$\beta = \sqrt[4]{\frac{3(1-\mu^2)}{36^2 \times 0.4^2}} = 0.3387$$

$$D = \frac{E(0.4)^3}{12(1-0.3^2)} = \frac{5.8608E}{1000}$$

$$\lambda = \sqrt[4]{\left(\frac{36.1}{0.2}\right)^2 3(1-0.3^2)} = 17.2695.$$

Thus Eq. (1) becomes

$$\frac{(200)(36)^2}{E(0.4)}(0.85) - \frac{H_o}{(2)(0.3387)^3(5.8608E/1000)} + \frac{M_o}{(2)(0.3387)^2(5.8608E/1000)}$$

$$= \frac{(200)(36.1)^2}{(2)(E)(0.2)}(1-0.3) + \frac{H_o(36.1)}{E(0.2)}(2 \times 17.2695) + \frac{2M_o(17.2695)^2}{E(0.2)}$$

$$550,800 - 2195.67H_o + 743.67M_o = 456,123.5 + 6234.29H_o + 2982.36M_o$$

$$M_o + 3.77H_o = 42.29. \tag{2}$$

The second compatibility equation at point a is given by rotation of cylinder due to $p + H_o + M_o$ = rotation of hemisphere due to $p + H_o + M_o$

$$0 - \frac{H_o}{(2)(0.3387)^2(5.8608E/1000)} + \frac{M_o}{(0.3387)(5.8608E/1000)}$$

$$= 0 - \frac{2H_o(17.2695)^2}{E(0.2)} - \frac{4M_o(17.2695)^3}{E(0.2)(36.1)}$$

$$1.50M_o + H_o = 0. \tag{3}$$

Solving Eqs. (2) and (3) yields

$$M_o = -9.08 \text{ inch-lbs/inch} \quad H_o = 13.63 \text{ lbs/inch}$$

The negative sign for the moment indicates that the actual moment is opposite that assumed in Figure 5.10b.

5.8.1 Cylindrical Shell

$$\text{Longitudinal bending stress} = \frac{6M_o}{t^2} = 340 \text{ psi}$$

$$\text{Longitudinal membrane stress} = \frac{pr}{2t} = 9000 \text{ psi}$$

$$\text{Total longitudinal stress} = 9340 \text{ psi}$$

From Eq. (3.76),

$$\text{Hoop bending stress} = 0.3 \times 340 = 100 \, \text{psi}$$

$$\text{Deflection at point } a = (550,800 - 2195.67H_o + 743.67M_o)/E$$

$$= 514,120/E.$$

The hoop membrane force is obtained from Eq. (3.72) as

$$N_\theta = \frac{Etw}{r} = 5712 \, \text{lbs/inch}$$

$$\text{Hoop membrane stress} = \frac{5712}{0.4} = 14,300 \, \text{psi}$$

$$\text{Total hoop stress} = 14,400 \, \text{psi}$$

5.8.2   Hemispherical Shell

$$\text{Longitudinal stress} = \frac{pR}{2t} + \frac{6M_o}{t^2}$$

$$= \frac{(200)(36.1)}{2(0.2)} + \frac{6(9.08)}{0.2^2}$$

$$= 19,410 \, \text{psi}$$

From Table 5.1,

$$\text{Hoop force} = \frac{pR}{2} + H_o\lambda - 2M_o\lambda^2/R$$

$$= \frac{(200)(36.1)}{2} + (13.63)(17.2695) - \frac{2(9.08)(17.2695)^2}{36.1}$$

$$= 18,480 \, \text{psi}$$

If a spherical shell has an axisymmetric hole and is subjected to edge loads, then Eq. (5.54) must be used to determine the constants K_1 through K_4. Other design functions are then established from the various equations derived. Table 5.2 lists various equations for open spherical shells.

5.9   Conical Shells

The derivation of the expressions for the bending moments in conical shells is obtained from the general equations of Section 5.7. In this case the angle is constant as shown in Figure 5.11.

Table 5.2 Edge loads in open spherical shells

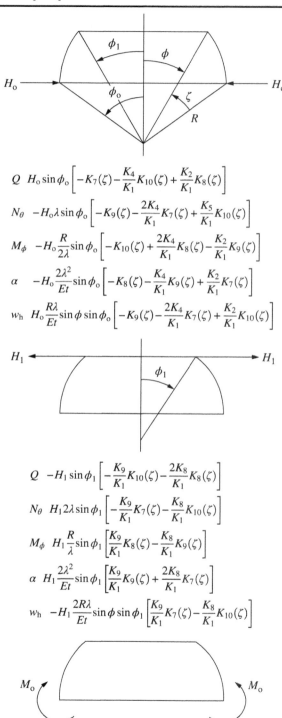

$$Q \quad H_o \sin\phi_o \left[-K_7(\zeta) - \frac{K_4}{K_1} K_{10}(\zeta) + \frac{K_2}{K_1} K_8(\zeta) \right]$$

$$N_\theta \quad -H_o \lambda \sin\phi_o \left[-K_9(\zeta) - \frac{2K_4}{K_1} K_7(\zeta) + \frac{K_5}{K_1} K_{10}(\zeta) \right]$$

$$M_\phi \quad -H_o \frac{R}{2\lambda} \sin\phi_o \left[-K_{10}(\zeta) + \frac{2K_4}{K_1} K_8(\zeta) - \frac{K_2}{K_1} K_9(\zeta) \right]$$

$$\alpha \quad -H_o \frac{2\lambda^2}{Et} \sin\phi_o \left[-K_8(\zeta) - \frac{K_4}{K_1} K_9(\zeta) + \frac{K_2}{K_1} K_7(\zeta) \right]$$

$$w_h \quad H_o \frac{R\lambda}{Et} \sin\phi \sin\phi_o \left[-K_9(\zeta) - \frac{2K_4}{K_1} K_7(\zeta) + \frac{K_2}{K_1} K_{10}(\zeta) \right]$$

$$Q \quad -H_1 \sin\phi_1 \left[-\frac{K_9}{K_1} K_{10}(\zeta) - \frac{2K_8}{K_1} K_8(\zeta) \right]$$

$$N_\theta \quad H_1 2\lambda \sin\phi_1 \left[-\frac{K_9}{K_1} K_7(\zeta) - \frac{K_8}{K_1} K_{10}(\zeta) \right]$$

$$M_\phi \quad H_1 \frac{R}{\lambda} \sin\phi_1 \left[\frac{K_9}{K_1} K_8(\zeta) - \frac{K_8}{K_1} K_9(\zeta) \right]$$

$$\alpha \quad H_1 \frac{2\lambda^2}{Et} \sin\phi_1 \left[\frac{K_9}{K_1} K_9(\zeta) + \frac{2K_8}{K_1} K_7(\zeta) \right]$$

$$w_h \quad -H_1 \frac{2R\lambda}{Et} \sin\phi \sin\phi_1 \left[\frac{K_9}{K_1} K_7(\zeta) - \frac{K_8}{K_1} K_{10}(\zeta) \right]$$

Table 5.2 *(continued)*

$$Q \quad -M_o\frac{2\lambda}{R}\left[\frac{K_6}{K_1}K_{13}(\zeta)+\frac{K_5}{K_1}K_{14}(\zeta)-\frac{K_3}{K_1}K_8(\zeta)\right]$$

$$N_\theta \quad M_o\frac{2\lambda^2}{R}\left[\frac{K_6}{K_1}K_{12}(\zeta)+\frac{K_5}{K_1}K_{11}(\zeta)-\frac{K_3}{K_1}K_{10}(\zeta)\right]$$

$$M_\phi \quad M_o\left[\frac{K_6}{K_1}K_{11}(\zeta)-\frac{K_5}{K_1}K_{12}(\zeta)+\frac{K_3}{K_1}K_9(\zeta)\right]$$

$$\alpha \quad M_o\frac{4\lambda^3}{EtR}\left[\frac{K_6}{K_1}K_{14}(\zeta)-\frac{K_5}{K_1}K_{13}(\zeta)-\frac{K_3}{K_1}K_7(\zeta)\right]$$

$$w_h \quad -M_o\frac{2\lambda^2}{Et}\sin\phi\left[\frac{K_6}{K_1}K_{12}(\zeta)+\frac{K_5}{K_1}K_{11}(\zeta)-\frac{K_3}{K_1}K_{10}(\zeta)\right]$$

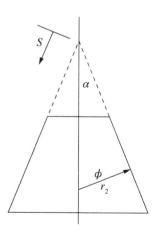

$M_1 \qquad\qquad\qquad M_1$

$$Q \quad -M_1\frac{2\lambda}{R}\left[-\frac{K_8}{K_1}K_{10}(\zeta)+\frac{K_{10}}{K_1}K_8(\zeta)\right]$$

$$N_\theta \quad M_1\frac{2\lambda^2}{R}\left[-\frac{2K_8}{K_1}K_7(\zeta)+\frac{K_{10}}{K_1}K_{10}(\zeta)\right]$$

$$M_\phi \quad M_1\left[\frac{2K_8}{K_1}K_8(\zeta)-\frac{K_{10}}{K_1}K_9(\zeta)\right]$$

$$\alpha \quad M_1\frac{4\lambda^3}{EtR}\left[\frac{K_8}{K_1}K_9(\zeta)+\frac{K_{10}}{K_1}K_7(\zeta)\right]$$

$$w_h \quad -M_1\frac{2\lambda^2}{Et}\sin\phi\left[-\frac{2K_8}{K_1}K_7(\zeta)+\frac{K_{10}}{K_1}K_{10}(\zeta)\right]$$

Source: Baker et al. (1968).
Notation: $M_\theta = \mu M_\phi$; $N_\phi = -Q\cot\phi$; w_h = horizontal component of deflection; α = rotation; $K_1 = \sinh^2\lambda - \sin^2\lambda$; $K_2 = \sinh^2\lambda + \sin^2\lambda$; $K_3 = \sinh\lambda\cosh\lambda + \sin\lambda\cos\lambda$; $K_4 = \sinh\lambda\cosh\lambda - \sin\lambda\cos\lambda$; $K_5 = \sin^2\lambda$; $K_6 = \sinh^2\lambda$; $K_7 = \cosh\lambda\cos\lambda$; $K_8 = \sinh\lambda\sin\lambda$; $K_8(\zeta) = \sinh\zeta\lambda\sin\zeta\lambda$; $K_9 = \cosh\lambda\sin\lambda - \sinh\lambda\cos\lambda$; $K_9(\zeta) = \cosh\zeta\lambda\sin\zeta\lambda - \sinh\zeta\lambda\cos\zeta\lambda$; $K_{10} = \cosh\lambda\sin\lambda + \sinh\lambda\cos\lambda$; $K_{10}(\zeta\lambda) = \cosh\zeta\lambda\sin\zeta\lambda + \sinh\zeta\lambda\cos\zeta\lambda$; $K_{11}(\zeta\lambda) = \cosh\zeta\lambda\cos\zeta\lambda - \sinh\zeta\lambda\sin\zeta\lambda$; $K_{12} = \cosh\lambda\cos\lambda + \sinh\lambda\sin\lambda$; $K_{12}(\zeta) = \cosh\zeta\lambda\cos\zeta\lambda + \sinh\zeta\lambda\sin\zeta\lambda$; $K_{13}(\zeta\lambda) = \cosh\zeta\lambda\sin\zeta\lambda$; $K_{14}(\zeta\lambda) = \sinh\zeta\lambda\cos\zeta\lambda$.

S

α

ϕ

r_2

Figure 5.11 Conical shell

Equations (5.26), (5.27), (5.28), (5.33), (5.34), (5.35), (5.41), and (5.42) have to be rewritten for conical shells with the following substitutions:

$$\phi = \frac{\pi}{2} - \alpha, \quad r_1 = \infty, \quad r_2 = s \tan \alpha$$

$$ds = r_1 d\phi.$$

The solutions of the resulting eight equations (Flugge 1967) involve Bessel functions. These solutions are too cumbersome to use on a regular basis. However, simplified asymptotic solutions, similar to those developed for spherical shells, can be developed for the large end of conical shells with $\beta > 11$ where

$$\beta = 2 \sqrt[4]{3(1-\mu^2)} \sqrt{\frac{s \cot \alpha}{t}}. \tag{5.64}$$

This range of β is common for most conical shells encountered in industry.

$$Q_s = \frac{e^\beta}{2^{3/4} s \sqrt{\pi\beta}} \left[C_1 \cos\left(\beta - \frac{\pi}{8}\right) + C_2 \sin\left(\beta - \frac{\pi}{8}\right) \right] \tag{5.65}$$

$$M_s = \frac{e^\beta}{\sqrt[4]{2}\beta\sqrt{\pi\beta}} \left[C_1 \sin\left(\beta + \frac{\pi}{8}\right) - C_2 \cos\left(\beta + \frac{\pi}{8}\right) \right] \tag{5.66}$$

$$M_\theta = \mu M_s \tag{5.67}$$

$$N_\theta = \frac{-\sqrt{\beta} e^\beta \tan \alpha}{2\sqrt[4]{2}\sqrt{\pi s}} \left[C_1 \cos\left(\beta + \frac{\pi}{8}\right) + C_2 \sin\left(\beta + \frac{\pi}{8}\right) \right] \tag{5.68}$$

$$N_s = Q_s \tan \alpha \tag{5.69}$$

$$w = \frac{s \sin \alpha}{Et} (N_\theta - \mu N_s) \tag{5.70}$$

where C_1 and C_2 are obtained from the boundary conditions.

Application of Eqs. (5.64) to (5.70) to edge forces of a full cone is given in Table 5.3. Table 5.4 lists various equations for truncated cones.

Example 5.8
Find the discontinuity forces in the cone Figure 5.12a, due to an internal pressure of 60 psi. Let $E = 30,000$ ksi and $\mu = 0.30$.

Solution
The deflection w at point A due to internal pressure, Figure 5.12b, is obtained from Eq. (3.95) by using the radius $r/\cos \alpha$ rather than r. Hence, radial deflection is expressed as

$$w = \frac{-pr^2}{Et\cos^2\alpha} (1 - \mu/2). \tag{1}$$

Table 5.3 Edge loads in conical shells (Jawad and Farr 1989)

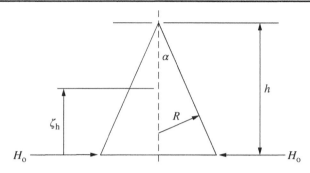

Edge functions $\zeta = 0$

$\theta \qquad -\dfrac{H_o h^2}{2D\beta^2 \cos\alpha}$

$w_h \qquad -\dfrac{H_o h^3}{2D\beta^3 \cos\alpha}\left[1 - \dfrac{\mu h \tan\alpha}{2R\beta\cos\alpha}\right]$

General functions

$Q \qquad \sqrt{2}H_o e^{-\beta\zeta}\cos\alpha\cos\left(\beta\zeta + \dfrac{\pi}{4}\right)$

$N_\theta \qquad -\dfrac{2H_o R^2 \beta\cos^2\alpha}{h}e^{-\beta\zeta}\cos\beta\zeta$

$M_\phi \qquad -\dfrac{H_o h}{\beta}e^{-\beta\zeta}\sin\beta\zeta$

$\theta \qquad -\dfrac{H_o h^2}{\sqrt{2}D\beta^2\cos\alpha}e^{-\beta\zeta}\sin\left(\beta\zeta + \dfrac{\pi}{4}\right)$

$w_h \qquad \dfrac{H_o h^3 e^{-\beta\zeta}}{2D\beta^3\cos\alpha}\left[\cos\beta\zeta - \dfrac{\mu h}{2R\beta}\tan\alpha\cos\left(\beta\zeta + \dfrac{\pi}{4}\right)\right]$

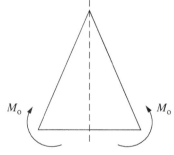

Edge functions $\zeta = 0$

$\theta \qquad \dfrac{M_o h}{D\beta\cos\alpha}$

$w_h \qquad -\dfrac{M_o h^2}{2D\beta^2\cos\alpha}$

(continued overleaf)

Table 5.3 *(continued)*

General functions

$$Q \quad -\frac{2M_o\beta}{h}\cos\alpha e^{-\beta\zeta}\sin\beta\zeta$$

$$N_\theta \quad -\frac{4R\beta^2 M_o}{\sqrt{2}h^2}\cos^2\alpha e^{-\beta\zeta}\cos\left(\beta\zeta+\frac{\pi}{4}\right)$$

$$M_\phi \quad \sqrt{2}M_o e^{-\beta\zeta}\sin\left(\beta\zeta+\frac{\pi}{4}\right)$$

$$\theta \quad \frac{M_o h}{D\beta\cos\alpha}e^{-\beta\zeta}\cos\beta\zeta$$

$$w_h \quad -\frac{M_o h^2}{2D\beta^2\cos\alpha}e^{-\beta\zeta}\left[\sqrt{2}\cos\left(\beta\zeta+\frac{\pi}{4}\right)+\frac{\mu h}{\beta R\cos\alpha}\tan\alpha\sin\beta\zeta\right]$$

Source: Jawad and Farr (1989). Reproduced with permission of John Wiley & Sons.
Notation: $M_\theta = \mu M_\phi$; $N_\phi = -Q\cot\alpha$; w_h = horizontal component of deflection; θ = rotation; $\beta = \left(h/\left(\sqrt{Rt}\cos\alpha\right)\right)\sqrt[4]{3(1-\mu^2)}$; μ = Poisson's ratio. Inward deflection is positive. Positive rotation is in direction of positive moments. Tensile N_ϕ and N_θ are positive. Positive moments cause tension on the inside surface. Inward Q is positive.

Table 5.4 Edge loads in open conical shells

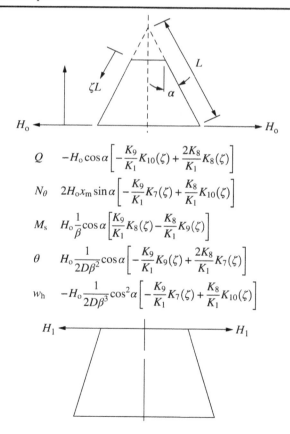

$$Q \quad -H_o\cos\alpha\left[-\frac{K_9}{K_1}K_{10}(\zeta)+\frac{2K_8}{K_1}K_8(\zeta)\right]$$

$$N_\theta \quad 2H_o x_m\sin\alpha\left[-\frac{K_9}{K_1}K_7(\zeta)+\frac{K_8}{K_1}K_{10}(\zeta)\right]$$

$$M_s \quad H_o\frac{1}{\beta}\cos\alpha\left[\frac{K_9}{K_1}K_8(\zeta)-\frac{K_8}{K_1}K_9(\zeta)\right]$$

$$\theta \quad H_o\frac{1}{2D\beta^2}\cos\alpha\left[-\frac{K_9}{K_1}K_9(\zeta)+\frac{2K_8}{K_1}K_7(\zeta)\right]$$

$$w_h \quad -H_o\frac{1}{2D\beta^3}\cos^2\alpha\left[-\frac{K_9}{K_1}K_7(\zeta)+\frac{K_8}{K_1}K_{10}(\zeta)\right]$$

Table 5.4 (*continued*)

$$Q \quad -H_1 \cos\alpha \left[K_7(\zeta) - \frac{K_4}{K_1} K_{10}(\zeta) + \frac{K_2}{K_1} K_8(\zeta) \right]$$

$$N_\theta \quad H_1 x_m \beta \sin\alpha \left[K_9(\zeta) + \frac{2K_4}{K_1} K_7(\zeta) - \frac{K_2}{K_1} K_{10}(\zeta) \right]$$

$$M_s \quad H_1 \frac{1}{2} \cos\alpha \left[K_{10}(\zeta) - \frac{2K_4}{K_1} K_8(\zeta) + \frac{K_2}{K_1} K_9(\zeta) \right]$$

$$\theta \quad H_1 \frac{\cos\alpha}{2D\beta^2} \left[-K_8(\zeta) + \frac{K_4}{K_1} K_9(\zeta) - \frac{K_2}{K_1} K_7(\zeta) \right]$$

$$w_h \quad -H_1 \frac{\cos^2\alpha}{4D\beta^3} \left[K_9(\zeta) + \frac{2K_4}{K_1} K_7(\zeta) - \frac{K_2}{K_1} K_{10}(\zeta) \right]$$

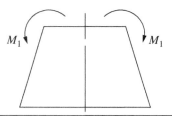

$$Q \quad -M_o 2\beta \left[\frac{K_8}{K_1} K_{10}(\zeta) - \frac{K_{10}}{K_1} K_8(\zeta) \right]$$

$$N_\theta \quad M_o 2\beta^2 x_m \tan\alpha \left[\frac{2K_8}{K_1} K_7(\zeta) - \frac{K_{10}}{K_1} K_{10}(\zeta) \right]$$

$$M_s \quad M_o \left[\frac{2K_8}{K_1} K_8(\zeta) - \frac{K_{10}}{K_1} K_9(\zeta) \right]$$

$$\theta \quad M_o \frac{1}{D\beta} \left[\frac{K_8}{K_1} K_9(\zeta) + \frac{K_{10}}{K_1} K_7(\zeta) \right]$$

$$w_h \quad -M_o \frac{\cos\alpha}{2D\beta^2} \left[\frac{2K_8}{K_1} K_7(\zeta) - \frac{K_{10}}{K_1} K_{10}(\zeta) \right]$$

(*continued overleaf*)

Table 5.4   (*continued*)

$$Q \quad -M_1 2\beta \left[\frac{K_6}{K_1}K_{13}(\zeta) + \frac{K_5}{K_1}K_{14}(\zeta) - \frac{K_3}{K_1}K_8(\zeta)\right]$$

$$N_\theta \quad M_1 2\beta^2 x_m \tan\alpha \left[\frac{K_6}{K_1}K_{12}(\zeta) + \frac{K_5}{K_1}K_{11}(\zeta) - \frac{K_3}{K_1}K_{10}(\zeta)\right]$$

$$M_s \quad M_1 \left[\frac{K_6}{K_1}K_{11}(\zeta) - \frac{K_5}{K_1}K_{12}(\zeta) + \frac{K_3}{K_1}K_9(\zeta)\right]$$

$$\theta \quad M_1 \frac{1}{D\beta} \left[\frac{K_6}{K_1}K_{14}(\zeta) - \frac{K_5}{K_1}K_{13}(\zeta) + \frac{K_3}{K_1}K_7(\zeta)\right]$$

$$w_h \quad -M_1 \frac{\cos\alpha}{2D\beta^2} \left[\frac{K_6}{K_1}K_{12}(\zeta) + \frac{K_5}{K_1}K_{11}(\zeta) - \frac{K_3}{K_1}K_{10}(\zeta)\right]$$

Source: Baker et al. (1968).
Notation: $M_\theta = \mu M_\phi$; $N_\theta = -Q \cot\alpha$; w_h = horizontal component of deflection; x_m = Slanted length measured from the apex to mid-distance of the truncated cone; θ = rotation; $\beta = \sqrt[4]{3(1-\mu^2)}/\sqrt{(tx_m \tan\alpha)}$; $K_1 = \sinh^2\beta L - \sin^2\beta L$; $K_2 = \sinh^2\beta L + \sin^2\beta L$; $K_3 = \sinh\beta L \cosh\beta L + \sin\beta L \cos\beta L$; $K_4 = \sinh\beta L \cosh\beta L - \sin\beta L \cos\beta L$; $K_5 = \sin^2\beta L$; $K_6 = \sinh^2\beta L$; $K_7 = \cosh\beta L \cos\beta L$; $K_7(\zeta) = \cosh\beta L\zeta \cos\beta L\zeta$; $K_8 = \sinh\beta L \sin\beta L$; $K_8(\zeta) = \sinh\beta L\zeta \sin\beta L\zeta$; $K_9 = \cosh\beta L \sin\beta L - \sinh\beta L \cos\beta L$; $K_9(\zeta) = \cosh\beta L\zeta \sin\beta L\zeta - \sinh\beta L\zeta \cos\beta L\zeta$; $K_{10} = \cosh\beta L \sin\beta L + \sinh\beta L \cos\beta L$; $K_{10}(\zeta) = \cosh\beta L\zeta \sin\beta L\zeta + \sinh\beta L\zeta \cos\beta L\zeta$; $K_{11}(\zeta) = \cosh\beta L\zeta \cos\beta L\zeta - \sinh\beta L\zeta \sin\beta L\zeta$; $K_{12}(\zeta) = \cosh\beta L\zeta \cos\beta L\zeta + \sinh\beta L\zeta \sin\beta L\zeta$; $K_{13}(\zeta) = \cosh\beta L\zeta \sin\beta L\zeta$; $K_{14}(\zeta) = \sinh\beta L\zeta \cos\beta L\zeta$.

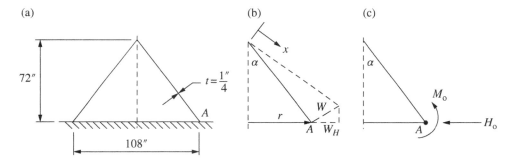

Figure 5.12   Cone with fixed edge: (a) conical shell, (b) deflection of cone, and (c) discontinuity forces

The horizontal deflection is

$$w_H = \frac{-pr^2}{Et\cos\alpha}(1-\mu/2)$$

$$= \frac{-(60)(54)^2(1-0.3/2)}{30,000,000(0.25)(0.8)}$$

$$w_H = -0.0248 \text{ inch.}$$

The rotation, θ, at point A is obtained by writing Eq. (1) as

$$w = \frac{-px^2\sin^2\alpha}{Et\cos^2\alpha}(1-\mu/2)$$

and from Eq. (5.33) with $\nu = 0$, and $dx = r_1\,d\phi$

$$\theta = -\frac{dw}{r_1 d\phi}$$

$$= -\frac{dw}{dx}$$

$$\theta = \frac{2px\sin^2\alpha}{Et\cos^2\alpha}(1-\mu/2)$$

$$= \frac{2pr\sin\alpha}{Et\cos^2\alpha}(1-\mu/2)$$

$$= \frac{2(60)(54)(0.6)(0.85)}{30,000,000(0.25)(0.64)}$$

$$\theta = 0.000689 \text{ rad.}$$

From Table 5.3,

$$R = r/\cos\alpha = 54/0.8$$

$$= 67.50$$

$$\beta = \frac{72}{\sqrt{67.50 \times 0.25}\ 0.60}\sqrt[4]{3(1-0.3^2)}$$

$$= 37.549$$

$$D = 42,926.$$

From Fig. 5.12c,

$$\text{Deflection due to } p + M_o + H_o = 0$$

or

$$-0.0248 - \frac{hM_o}{2D\beta^2\cos\alpha} + \frac{h^3 H_o}{2D\beta^3\cos\alpha}\left(1 - \frac{\mu h\tan\alpha}{2R\beta\cos\alpha}\right) = 0. \tag{2}$$

$$M_o - 1.910\,H_o = -463.21$$

Similarly,

$$\text{Rotation due to } p + M_o + H_o = 0$$

$$-0.000689 - \frac{hM_o}{D\beta\cos\alpha} + \frac{h^2 H_o}{2D\beta^2\cos\alpha} = 0$$

or

$$M_o - 0.959 H_o = -12.339. \tag{3}$$

Solving Eqs. (2) and (3) results in

$$M_o = 442.3 \text{ inch-lbs/inch.}$$
$$H_o = 474.1 \text{ lbs/inch.}$$

Problems

5.1 Calculate the required thickness of a spherical head subjected to internal pressure of 4000 psi using elastic analysis, Eq. (5.6); plastic analysis, Eq. (5.20); and creep analysis, Eq. (5.21). Let inside radius $R = 24$ inch, allowable stress $S = 20{,}000$ psi, and $n = 6.0$.

5.2 Determine the discontinuity stress at the spherical-to-cylindrical junction shown. The dimension of the stiffening ring at the junction is 4 inch by 3/4 inch. Let $E = 30{,}000$ ksi and $\mu = 0.3$.

5.3 Determine the length $L = \zeta R$ where the moment diminishes to 1% of the moment M_o applied at the free edge. How does this length compare with that in Example 3.5 for cylindrical shells?

5.4 Derive Eq. (5.63).

5.5 A conical shell is welded to a cylindrical shell as shown.

1. Calculate the thickness of the cylindrical and conical shells.
2. Calculate the required area of the stiffening ring at the junction.
3. Calculate the discontinuity stresses at the junction.

Let internal pressure $P = 450$ psi, allowable stress for cylindrical and conical shells $S = 20{,}000$ psi, allowable stress for stiffening ring $= 10{,}000$ psi, $E = 27{,}000$ ksi, $\propto = 450$, and $\mu = 0.3$.

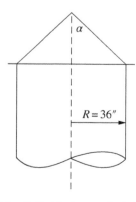

Problem 5.5   Cylindrical to conical shell junction

6

Buckling of Shells of Revolution

6.1 Elastic Buckling of Spherical Shells

The buckling of spherical shells under external pressure has been investigated by numerous researchers. Von Karman and Tsien (1939) developed a solution that fits experimental data very closely. Taking a buckled section of a spherical shell, Figure 6.1, he made the following assumptions:

1. The deflected shape is rotationally symmetric.
2. The buckled length is small.
3. The deflection of any element of the shell is parallel to axis of rotational symmetry.
4. The effect of lateral contraction due to Poisson's ratio is neglected.

Based on these assumptions and Figure 6.1, the strain due to extension of the element is

$$\varepsilon = \frac{(dr/\cos\theta)-(dr/\cos\alpha)}{dr/\cos\alpha}$$

$$\varepsilon = \frac{\cos\alpha}{\cos\theta} - 1.$$

The strain energy of the extension of the element is

$$U_1 = \frac{ER^3}{2}\left(\frac{t}{R}\right)2\pi\int_0^\beta \left(\frac{\cos\alpha}{\cos\theta}-1\right)^2 \sin\alpha\, d\alpha. \tag{6.1}$$

Stress in ASME Pressure Vessels, Boilers, and Nuclear Components, First Edition. Maan H. Jawad.
© 2018, The American Society of Mechanical Engineers (ASME), 2 Park Avenue,
New York, NY, 10016, USA (www.asme.org). Published 2018 by John Wiley & Sons, Inc.

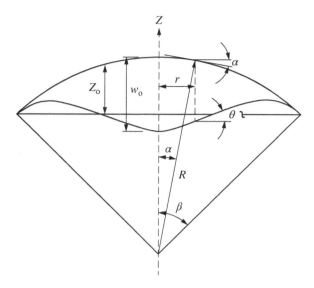

Figure 6.1   Buckled shape of a spherical segment

The strain energy due to compression of the shell prior to buckling is

$$U_2 = \frac{-ER^3}{2}\left(\frac{t}{R}\right)2\pi\int_0^\beta \left(\frac{pR}{2Et}\right)^2 \alpha\, d\alpha. \tag{6.2}$$

Similarly, the strain energy due to bending is given by

$$U_3 = \frac{ER^3}{2}\left(\frac{t}{R}\right)^3 \frac{\pi}{6}\int_0^\beta \sin\alpha\left[\left(\frac{\cos\theta\, d\theta}{\cos\alpha\, d\alpha}-1\right)^2 + \left(\frac{\sin\theta}{\sin\alpha}-1\right)^2\right]d\alpha. \tag{6.3}$$

The external work is equal to the applied pressure times the volume included between the deflected and original surfaces. This can be expressed as

$$W = R^3\pi\int_0^\beta \sin^2\alpha(\tan\theta-\tan\alpha)\cos\alpha\, d\alpha. \tag{6.4}$$

The total potential energy, $\prod$, of the system is given by

$$\prod = U_1 + U_2 + U_3 - W. \tag{6.5}$$

The terms in Eq. (6.5) can be simplified substantially by assuming β to be small. Then, expanding the sine and cosine expressions into a power series and neglecting terms higher than the third order, Eq. (6.5) becomes

$$\Pi = E\left(\frac{t}{R}\right)\int_0^\beta \left[\frac{1}{2}(\theta^2 - \alpha^2) - \frac{PR}{2Et}\right]^2 \alpha\, d\alpha$$

$$+ \frac{E(t/R)^3}{12}\int_0^\beta \left[\left(\frac{d\theta}{d\alpha} - 1\right)^2 + \left(\frac{\theta}{\alpha} - 1\right)^2\right]\alpha\, d\alpha + p\int_0^\beta \alpha^2(\theta - \alpha)d\alpha.$$

(6.6)

Minimizing this equation with respect to θ results in an equation that is too cumbersome to solve. Accordingly, Von Karman assumed an expression for θ that satisfies the boundary condition $\theta = 0$ at $\alpha = 0$ and $\theta = \beta$ at $\alpha = \beta$, in the form

$$\theta = \left\{1 - K\left[1 - \left(\frac{\alpha}{\beta}\right)^2\right]\right\}$$

(6.7)

where K is a constant.

Substituting Eq. (6.7) into the expression

$$\frac{\partial \Pi}{\partial K} = 0$$

and utilizing the relationship $\sigma = pR/2t$, we get

$$\frac{\sigma}{E} = \frac{1}{70}\beta^2\left(28 - 21K + 4K^2\right) + \frac{4}{3}\left(\frac{t}{R}\right)^2\frac{1}{\beta^2}.$$

(6.8)

The ordinate Z_o in Figure 6.1 is calculated from

$$Z_o + \int_0^\beta \left(\frac{dz}{dr}\right)dr = 0$$

or

$$Z_o = R\int_0^\beta \tan\theta\cos\alpha\, d\alpha.$$

The ordinate of the shell at the center before deflection is given by $R(1 - \cos\beta)$. The deflection, w_o, at the center is then expressed by

$$w_o = R\int_0^\beta (\tan\alpha - \tan\theta)\cos\alpha\, d\alpha$$

or, assuming β to be small,

$$w_o = R\int_0^\beta (\alpha - \theta)\, d\alpha.$$

(6.9)

Substituting Eq. (6.9) into Eq. (6.7) gives

$$K = \frac{4w_o}{R\beta^2}.$$

(6.10)

Substituting Eq. (6.10) into Eq. (6.8) results in

$$\frac{\sigma}{E} = \frac{2}{5}\beta^2 - \frac{6w_o}{5R} + \left[\frac{32}{35}\frac{w_o^2}{R^2} + \frac{4t^2}{3R^2}\right]\frac{1}{\beta^2}.$$

(6.11)

The minimum value of Eq. (6.11) is obtained by differentiating with respect to β^2 and equating to zero. This gives

$$\beta^2 = \sqrt{\frac{16}{7}\left(\frac{w_o}{R}\right)^2 + \frac{10}{3}\left(\frac{t}{R}\right)^2}$$

and Eq. (6.11) becomes

$$\frac{\sigma R}{Et} = \frac{4}{5}\left[\sqrt{\frac{16}{7}\left(\frac{w_o}{t}\right)^2 + \frac{10}{3}} - \frac{3}{2}\left(\frac{w_o}{t}\right)\right].$$

(6.12)

A plot of this equation is shown in Figure 6.2. The minimum buckling value is

$$\sigma_{cr} = 0.183\, Et/R.$$

(6.13)

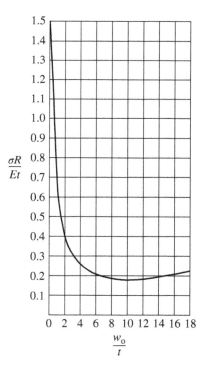

Figure 6.2   A plot of Eq. (6.12)

6.2 ASME Procedure for External Pressure

The ASME pressure vessel code Sections VIII-1 and III-NC use Eq. (6.13) as a basis for design in the elastic range. This equation can be written as

$$\varepsilon_{cr} = \frac{\sigma_{cr}}{E} = \frac{0.183}{R/T}.$$

$$(6.14)$$

Equation (6.14) is modified by ASME by using a knockdown factor of 1.5 to account for the effect of geometric imperfections on external pressure calculation. The equation for strain is expressed as

$$\varepsilon = \frac{0.125}{R/t}.$$

$$(6.15)$$

In the elastic range the design pressure is obtained from Eq. (6.14) by using the quantity $\sigma = PR/2t$ with a knockdown factor of 1.5 and a design factor of 4.0. The design equation in the elastic range becomes

$$P = 0.0625E/(R/t)^2.$$

$$(6.16)$$

In the plastic range the value of E in Eq. (6.13) can conservatively be replaced by tangent modulus E_t, and an external pressure chart may be used similar to that in Figure 4.15. In this case Eq. (6.15) is written as

$$A = \frac{0.125}{R/t}.$$

$$(6.17)$$

And the equation for pressure using a design factor of four is

$$P = \frac{B}{R/t}$$

$$(6.18)$$

where B is obtained from Figure 4.15 and is equal to $\sigma/2$.

Example 6.1
What is the allowable external pressure on an ASME hemispherical head with $R = 72$ inch, $t = 0.25$ inch, and $E = 29,000$ ksi?

Solution
From Eq. (6.17), $A = 0.125/(72/0.25) = 0.00043$.
From Figure 4.15, $B \approx 6400$ psi and is in the elastic region.
From Eq. (6.18), $P = 6400/(72/0.25) = 22.2$ psi.

Since B is in the elastic region, we can calculate the value from Eq. (6.16):

$$P = (0.0625)(29,000)/(72/0.25)^2 = 21.9\,\text{psi}$$

6.3   Buckling of Stiffened Spherical Shells

Equation (6.13) can be written in terms of buckling pressure as

$$p_{cr} = 0.366E(t/R)^2 \tag{6.19}$$

or

$$t = R\sqrt{\frac{p_{cr}}{0.366E}}. \tag{6.20}$$

Equation (6.20) shows that for a given external pressure and modulus of elasticity, the required thickness is proportional to the radius of the spherical section. As the radius gets larger, so does the thickness. One procedure for reducing the membrane thickness is by utilizing stiffened shells. Thus, for the shell shown in Figure 6.3 with closely spaced stiffeners, the buckling pressure (Buchert 1964) is obtained by modifying Eq. (6.19) as

$$p_{cr} = 0.366E\left(\frac{t_m}{R}\right)^2\left(\frac{t_b}{t_m}\right)^{3/2} \tag{6.21}$$

in which
  t_m = effective membrane thickness

$$= t + \left(\frac{A}{d}\right)$$

t_b = effective bending thickness

$$= \left(\frac{12I}{d}\right)^{1/3}$$

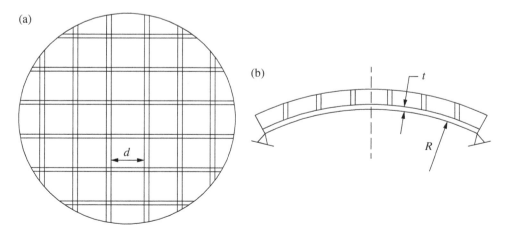

Figure 6.3   Stiffened spherical shell: (a) plan view and (b) cross section

where

A = area of stiffening ring
d = spacing between stiffeners
I = moment of inertia of ring
t = thickness of shell

For large spherical structures such as large tank roofs and stadium domes, the stiffener spacing in Figure 6.3 increases significantly. In this case, the composite buckling strength of shell and stiffeners (Buchert 1966) is expressed by

$$p_{cr} = 0.366E \left(\frac{t}{R}\right)^2 \left(1 + \frac{12I}{dt^3}\right)^{1/2} \left(1 + \frac{A}{dt}\right)^{1/2}. \tag{6.22}$$

Local buckling of the shell between the stiffeners must also be considered for large-diameter shells. One such equation is given by

$$p_{cr} = 7.42 \frac{Et^3}{Rd^2}. \tag{6.23}$$

It should be noted that large edge rotations and deflections can reduce the value obtained by Eq. (6.22) significantly.

Equation (6.22) is also based on the assumption that the spacing of the stiffeners is the same in the circumferential and meridional directions. Other equations can be developed (Buchert 1966) for unequal spacing of stiffeners.

6.4 Ellipsoidal Shells

The allowable external pressure for ellipsoidal shells is calculated using the procedure developed previously for spherical shells. The applicable spherical radius is that at the apex of the ellipsoidal shell.

6.5 Buckling of Conical Shells

The derivation of the equations for the buckling of conical shells is fairly complicated and beyond the scope of this book. The derivation (Niordson 1947) for the buckling pressure of the cone shown in Figure 6.4 consists of obtaining expressions for the work done by the applied pressure, membrane forces, stretching of the middle surface, and bending of the cone. The total work is then minimized to obtain a critical pressure expression in the form

$$p_{cr} = \frac{Et}{\rho_o \, a_o^2} f \frac{t^2 \, a_o^4}{(1-\mu^2) \, \rho_o^2} \tag{6.24}$$

where

E = modulus of elasticity
t = thickness of cone

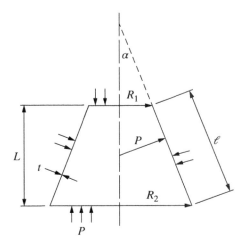

Figure 6.4 Conical section

$a_o = \lambda \rho (1 - \beta/2)$

$\lambda = \pi/\ell$

$\rho_o = \rho(1 - \beta/2)$

$\beta = \dfrac{\ell}{\rho} \tan \alpha$

Equation (6.24) is very cumbersome to use due to the iterative process required. Seide (1962) showed that Eq. (6.24) for the buckling of a conical shell is similar to the equation for the buckling of a cylindrical shell having a length equal to the slant length of the cone and a radius equal to the average radius of the cone. He also showed that the buckling of a cone is affected by the function $f(1 - R_1/R_2)$ and is expressed as

$$p_{cr} = \bar{p} f (1 - R_1/R_2) \tag{6.25}$$

where, $\bar{p}$ = pressure of equivalent cylinder as defined previously and f = cone function as defined in Figure 6.5.

By various substitutions (Jawad 1980), it can be shown that Eq. (6.25) can be transferred to the form of Eq. (4.65) as

$$p_{cr} = \frac{0.92E(t_e/R_2)^{2.5}}{L_e/R_2} \text{for metallic cones} \tag{6.26}$$

where

$$t_e = \text{effective thickness of cone}$$

$$= t \cos \alpha \tag{6.27}$$

$$t = \text{thickness of cone}$$

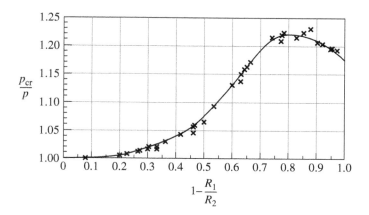

Figure 6.5   Plot of Eq. (6.25). Source: Jawad (1980). Reproduced with permission of ASME

and

$$L_e = \text{effective length of cone}$$

$$= \frac{L}{2}(1 + R_1/R_2).$$ (6.28)

Thus, conical shells subjected to external pressure may be analyzed as cylindrical shells with an effective thickness and length as defined by Eqs. (6.27) and (6.28).

Example 6.2
Determine the allowable external pressure, at room temperature, for the steel cone shown in Figure 6.6 by using (a) Eq. (6.26) and (b) Eq. (4.63). Let F.S. = 3.0 and $E = 29{,}000$ ksi.

Solution
a. From Figure 6.6, $\alpha = 17.35°$.
   From Eq. (6.27), $t_e = (3/32)(\cos 17.35) = 0.0895$ inch.
   From Eq. (6.28), $L_e = \dfrac{32}{2}\left(1 + \dfrac{20}{30}\right) = 26.67$ inch.
   Hence, from Eq. (6.26),

$$p = \frac{0.92 \times 29{,}000{,}000(0.0895/30)^{2.5}}{(26.67/30)(3)}$$

$$= 4.86\,\text{psi}.$$

b. $\dfrac{L_e}{D_2} = \dfrac{26.67}{60} = 0.44.$

   $\dfrac{D_2}{t_e} = \dfrac{60}{0.0895} = 670.$

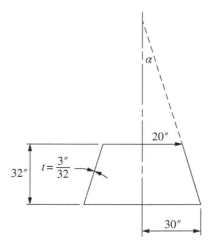

Figure 6.6   Steel conical section

From Figure 4.14, $A = 0.00018\,\text{inch/inch}$.

From Figure 4.15, $B = 2500\,\text{psi}$.

and from Eq. (4.63),

$$p = \frac{(2)(0.00018)(29,000,000)}{3(670)}$$

$$= 5.19\,\text{psi}.$$

It should be remembered that there is an advantage in using method (b) in that the actual stress–strain diagram of the material is used. This takes into account the plastic region if σ falls in that range.

6.6   Various Shapes

Further theoretical and experimental research is still needed to establish buckling strength of various configurations and shapes. This includes eccentric cones, torispherical shells, toriconical shells, and stiffened shells. Additionally, the effect of out-of-roundness and edge conditions on the buckling strength of shells needs further investigation.

Problems

6.1 Find the required thickness of the hemispherical shell in Problem 4.2. Use a factor of safety (F.S.) of 3.0.

6.2 A spherical aluminum diving chamber is under 1000 ft of water. Determine the required thickness. Let $R = 36$ inches, F.S. = 1.5, and $E = 10,000$ ksi.

6.3 Determine the allowable external pressure on the stiffened head in Figure 6.3. Assume $R = 13$ ft, $t = 3/8$ inch, $d = 8$ inches, size of stiffeners is 4 inch × 3/8 inch, $E = 29,000$ ksi, and a F.S. of 10.

6.4 Use Eqs. (6.22) and (6.23) to determine the required thickness of dome roof and the size and spacing of stiffeners. Let $R = 200$ ft, $E = 29,000$ ksi, $p = 85$ psf, and F.S. = 10.

6.5 Determine the required thickness of the cone shown in Problem 5.5 due to full vacuum. Use (a) Eq. (6.26) and (b) Eq. (4.62). Let $E = 29,000$ ksi and F.S. = 3.0.

6.6 Determine the required thickness of the cone shown in Figure 6.6 due to full vacuum. Let $E = 29,000$ ksi and use F.S. = 3.0.

7

Bending of Rectangular Plates

7.1 Introduction

Many structures such as boiler casings (Figure 7.1), submarine bulkheads, ship and barge hulls, ducts (Figure 7.2), aircraft components, and machine parts are designed in accordance with the general bending theory of plates. In this chapter, the pertinent equations of the bending theory of plates are developed, and examples are solved to demonstrate its applicability. These equations are applicable to thin plates subjected to small deflections. The majority of industrial applications encountered by the engineer fall under this category. Other theories dealing with thin plates with large deflections such as diaphragms, thick plates such as some tubesheets in heat exchangers, and composite and laminated plates in some aircraft components are beyond the scope of this book. Many of the references given at the end of the book delve into such theories, and the interested reader is encouraged to read them.

The derivation of the pertinent equations for the bending of thin plates with small deflections is based on the following assumptions:

1. The plate is assumed thin; that is, the plate thickness is substantially less than the lateral dimensions. An approximate rule of thumb is that the ratio of thickness to lateral dimension is of the order of about 0.1. This assumption is made in order to develop simplified equations applicable to a large variety of plates. As the ratio of thickness to lateral dimension increases, the stresses calculated by the thin plate theory become less conservative due to increased effect of the shearing forces, which are neglected in thin plate theory (Timoshenko and Woinowsky-Krieger 1959).
2. Applied loads are perpendicular to the middle surface of the plate. This assumption eliminates the need to consider in-plane membrane forces that are not considered in the classical theory of bending of thin plates. In-plane forces will be discussed briefly in Chapter 10 that deals with buckling of plates.

Stress in ASME Pressure Vessels, Boilers, and Nuclear Components, First Edition. Maan H. Jawad.
© 2018, The American Society of Mechanical Engineers (ASME), 2 Park Avenue,
New York, NY, 10016, USA (www.asme.org). Published 2018 by John Wiley & Sons, Inc.

Figure 7.1   Boiler casing. Source: Courtesy of the Nooter Corporation, St. Louis, MO.

3. The plate undergoes small deflections due to applied loads. Small deflections are defined as being less than the thickness of the plate. An approximate rule of thumb for defining small deflections is that the ratio of deflection to thickness is less than about 0.5. As this ratio increased above 0.5, the equations generally become overly conservative due to the introduction of membrane forces in the plate that tend to reduce the bending stresses predicted by the small deflection plate theory (Timoshenko and Woinowsky-Krieger 1959).

4. The undeflected middle surface of the plate coincides with the chosen x-y plane as shown in Figure 7.3. Points on the middle surface undergoing small deflections due to transverse loads on the plate are assumed to move perpendicular to the x-y plane to form a new middle surface.

5. Cross-sectional planes perpendicular to the middle surface prior to deflection remain straight and perpendicular to the middle surface subsequent to its deflection. This implies that the middle surface does not undergo any extension and that the middle surface is also the neutral surface of the plate.

6. The plate material is isotropic and homogeneous. This assumption results in equations that are greatly simplified and easy to apply to a vast variety of plate components.

Figure 7.2   Duct structure. Source: Courtesy of the Nooter Corporation, St. Louis, MO.

7.2   Strain–Deflection Equations

The relationship between strain and deflection of a thin plate is available from geometric considerations. We begin the derivation by letting the infinitesimal section in Figure 7.3 undergo some bending deformation. Downward deformation is defined as positive. The change in length at a distance z from the middle surface is obtained from the figure as

$$\frac{dx}{r_x} = \frac{dx + \varepsilon_x dx}{r_x + z} \tag{7.1}$$

where

dx = infinitesimal length in the x-direction
ε_x = strain in the x-direction

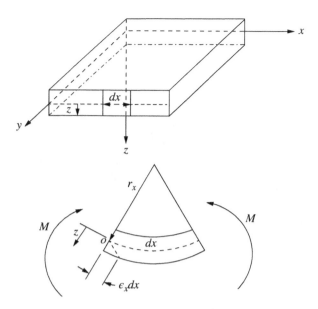

Figure 7.3 Bending of a rectangular plate

r_x = radius of curvature of the deformed middle surface in the x-direction
z = distance from the middle surface

Equation (7.1) may be simplified as

$$\varepsilon_x = \frac{z}{r_x}. \tag{7.2}$$

The curvature χ_x of the deformed middle surface is related to the radius of curvature by the relationship $\chi_x = 1/r_x$. Curvature is assumed positive if it is convex downward. Equation (7.2) can then be written as

$$\varepsilon_x = \chi_x z. \tag{7.3}$$

Similarly, in the y-direction,

$$\varepsilon_y = \frac{z}{r_y} \tag{7.4}$$

and

$$\varepsilon_y = \chi_y z \tag{7.5}$$

where

ε_y = strain in the y-direction
r_y = radius of curvature of the deformed middle surface in the y-direction
χ_y = curvature in the y-direction

The quantity χ_x is related to the deflection w and slope dw/dx by the expression (Shenk 1997)

$$\chi_x = -\frac{d^2w}{dx^2}\left[1 + \left(\frac{dw}{dx}\right)^2\right]^{-3/2}.$$

The expression d^2w/dx^2 in the nominator may be written as $(d/dx)(dw/dx)$. The negative sign in the nominator indicates that the slope dw/dx is decreasing as the location moves away from the origin point o in Figure 7.3.

For a small deflection w, the square of the slope dw/dx in the denominator is small compared to the quantity 1.0 and can thus be neglected. The previous expression becomes

$$\chi_x = -\frac{d^2w}{dx^2}. \tag{7.6}$$

Similarly, in the y-direction,

$$\chi_y = -\frac{d^2w}{dy^2}. \tag{7.7}$$

Substituting Eqs. (7.6) and (7.7) into Eqs. (7.3) and (7.5) gives the relationship between strain and deflection:

$$\varepsilon_x = -z\frac{d^2w}{dx^2} \tag{7.8}$$

$$\varepsilon_y = -z\frac{d^2w}{dy^2}. \tag{7.9}$$

The shearing strain–deformation relationship is obtained from Figure 7.4. If an infinitesimal element of length dx and width dy undergoes shearing deformations, α and β, due to in-plane shearing forces and twisting moments, then from Figure 7.4a

$$\sin\alpha \approx \alpha \approx \frac{\frac{\partial u}{\partial y}dy}{\left(1 + \frac{\partial v}{\partial y}\right)dy}.$$

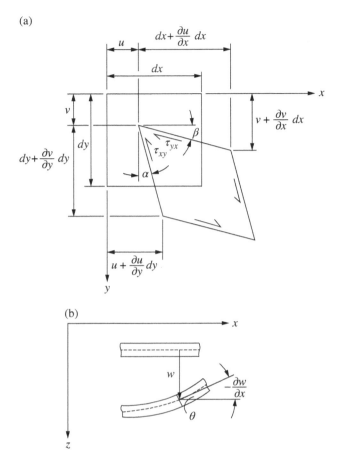

Figure 7.4 Shear deformation of a rectangular plate: (a) shear of an infinitesimal element and (b) deflection of element

or, for small shearing angles,

$$\alpha = \frac{\partial u}{\partial y}.$$

Similarly,

$$\sin\beta \approx \beta \approx \frac{\dfrac{\partial v}{\partial x}dx}{\left(1+\dfrac{\partial u}{\partial x}\right)dx}$$

$$\beta = \frac{\partial v}{\partial x}.$$

Hence,

$$\gamma_{xy} = \alpha + \beta = \frac{\partial u}{\partial y} + \frac{\partial v}{\partial x} \tag{7.10}$$

where

u = deflection in the x-direction
v = deflection in the y-direction
γ_{xy} = shearing strain
$\dfrac{\partial u}{\partial y}, \dfrac{\partial v}{\partial x}$ = shearing strains due to twisting

The rotation of the middle surface is shown in Figure 7.4b and is given by $\partial w/\partial x$. Due to this rotation, any point at distance z from the middle surface will deflect by the amount

$$u = z \ \tan \theta \approx z\theta$$

or

$$u = -z\frac{\partial w}{\partial x} \quad v = -z\frac{\partial w}{\partial y}.$$

Hence, Eq. (7.10) becomes

$$\gamma_{xy} = -2z\frac{\partial^2 w}{\partial x\, \partial y}. \tag{7.11}$$

Equations (7.8), (7.9), and (7.11) can be written as

$$\begin{bmatrix} \varepsilon_x \\ \varepsilon_y \\ \gamma_{xy} \end{bmatrix} = -z \begin{bmatrix} 1 & 0 & 0 \\ 0 & 1 & 0 \\ 0 & 0 & 2 \end{bmatrix} \begin{bmatrix} \dfrac{\partial^2 w}{\partial x^2} \\[2ex] \dfrac{\partial^2 w}{\partial y^2} \\[2ex] \dfrac{\partial^2 w}{\partial x\, \partial y} \end{bmatrix} \tag{7.12}$$

and are sufficiently accurate for developing the bending theory of thin plates. More precise strain expressions that are a function of the three displacement functions u, v, and w will be derived later when the buckling theory of thin plates is discussed.

7.3   Stress–Deflection Expressions

Equation (7.12) can be expressed in terms of stress rather than strain for ease of calculation. The relationship between stress and strain, excluding thermal loads, in a three-dimensional homogeneous and isotropic element (Figure 7.5) is obtained from the theory of elasticity (Sokolnikoff 1956) as

$$
\begin{bmatrix} \varepsilon_x \\ \varepsilon_y \\ \varepsilon_z \\ \gamma_{xy} \\ \gamma_{yz} \\ \gamma_{zx} \end{bmatrix} = \frac{1}{E}
\begin{bmatrix}
1 & -\mu & -\mu & 0 & 0 & 0 \\
-\mu & 1 & -\mu & 0 & 0 & 0 \\
-\mu & -\mu & 1 & 0 & 0 & 0 \\
0 & 0 & 0 & 2(1+\mu) & 0 & 0 \\
0 & 0 & 0 & 0 & 2(1+\mu) & 0 \\
0 & 0 & 0 & 0 & 0 & 2(1+\mu)
\end{bmatrix}
\begin{bmatrix} \sigma_x \\ \sigma_y \\ \sigma_z \\ \tau_{xy} \\ \tau_{yz} \\ \tau_{zx} \end{bmatrix}
\tag{7.13}
$$

where

$\varepsilon =$ axial strain
$\sigma =$ axial stress
$\gamma =$ shearing strain
$\tau =$ shearing stress
$E =$ modulus of elasticity
$\mu =$ Poisson's ratio

The quantities x, y, and z refer to the directions shown in Figure 7.5. The quantity $2(1+\mu)/E$ is usually written as $1/G$ where G is called the shearing modulus.

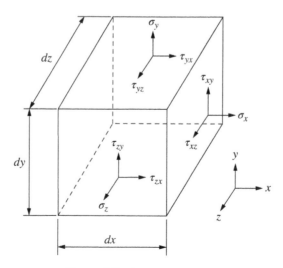

Figure 7.5   Stress components

The stress perpendicular to the surface, that is, in the z-direction, has a maximum value equal to the applied pressure. For the majority of plate applications in bending, the stress σ_z in the z-direction is small compared with the stress in the other two directions and thus can be neglected. In addition, the shearing stresses τ_{yz} and τ_{zx} are not needed in the formulation of a two-dimensional state of stress. Hence, for this condition, Eq. (7.13) can be written as

$$\begin{bmatrix} \sigma_x \\ \sigma_y \\ \tau_{xy} \end{bmatrix} = \frac{E}{1-\mu^2} \begin{bmatrix} 1 & \mu & 0 \\ \mu & 1 & 0 \\ 0 & 0 & \frac{1-\mu}{2} \end{bmatrix} \begin{bmatrix} \varepsilon_x \\ \varepsilon_y \\ \gamma_{xy} \end{bmatrix}. \tag{7.14}$$

Substituting Eq. (7.12) into Eq. (7.14) gives

$$\begin{bmatrix} \sigma_x \\ \sigma_y \\ \tau_{xy} \end{bmatrix} = \frac{-Ez}{1-\mu^2} \begin{bmatrix} 1 & \mu & 0 \\ \mu & 1 & 0 \\ 0 & 0 & (1-\mu) \end{bmatrix} \begin{bmatrix} \dfrac{\partial^2 w}{\partial x^2} \\[2ex] \dfrac{\partial^2 w}{\partial y^2} \\[2ex] \dfrac{\partial^2 w}{\partial x\,\partial y} \end{bmatrix}. \tag{7.15}$$

The elastic moduli of elasticity and Poisson's ratio for some commonly used materials are given in Table 7.1. The value of Poisson's ratio is relatively constant at various temperatures for a given material and is thus listed only for room temperature in Table 7.1.

Table 7.1 Moduli of elasticity and Poisson's ratio

Material	Poisson's ratio	Room temperature	Modulus of elasticity[a]				
			Temperature, °F				
			200	400	600	800	1000
Aluminum (6061)	0.33	10.0	9.6	8.7			
Brass (C71000)	0.33	20.0	19.5	18.8	17.8		
Bronze (C61400)	0.33	17.0	16.6	16.0	15.1		
Carbon steel ($C < 0.3$)	0.29	29.5	28.8	27.7	26.7	24.2	20.1
Copper (C12300)	0.33	17.0	16.6	16.0	15.1		
Cu–Ni (70–30) (C71500)	0.33	22.0	21.5	20.7	19.6		
Nickel alloy C276	0.29	29.8	29.1	28.3	27.6	26.5	25.3
Nickel alloy 600	0.29	31.0	30.2	29.5	28.7	27.6	26.4
Stainless steel (304)	0.31	28.3	27.6	26.5	25.3	24.1	22.8
Titanium (Gr. 1,2)	0.32	15.5	15.0	14.0	12.6	11.2	
Zirconium alloys	0.35	14.4	13.4	11.5	9.9		

[a] In million psi.

7.4 Force–Stress Expressions

Equation (7.15) can be utilized more readily when the stress values are replaced by moments. This is because the moments at the edges of the plate are needed to satisfy some of the boundary conditions in solving the differential equation. The relationship between moment and stress is obtained from Figure 7.6a. The moments shown in Figure 7.6b are positive and are per unit length. By definition, the sum of the moments about the neutral axis due to the internal forces is equal to the sum of the moments of the external forces. Hence,

$$
\begin{bmatrix} M_x \\ M_y \\ -M_{xy} \end{bmatrix} = \int_{-t/2}^{t/2} \begin{bmatrix} \sigma_x \\ \sigma_x \\ \tau_{xy} \end{bmatrix} z\,dz.
\tag{7.16}
$$

The negative sign of M_{xy} in Eq. (7.16) is needed since the direction of M_{xy} in Figure 7.6b results in a shearing stress τ_{xy} that has a direction opposite to that defined in Figure 7.4a in the positive z-axis. Substituting Eq. (7.15) into Eq. (7.16) results in

$$
\begin{bmatrix} M_x \\ M_y \\ M_{xy} \end{bmatrix} = -D \begin{bmatrix} 1 & \mu & 0 \\ \mu & 1 & 0 \\ 0 & 0 & -(1-\mu) \end{bmatrix} \begin{bmatrix} \dfrac{\partial^2 w}{\partial x^2} \\ \dfrac{\partial^2 w}{\partial y^2} \\ \dfrac{\partial^2 w}{\partial x\,\partial y} \end{bmatrix}
\tag{7.17}
$$

where

$$
D = \frac{Et^3}{12(1-\mu^2)}
\tag{7.18}
$$

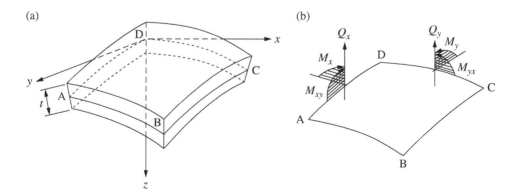

Figure 7.6   Moments in a plate: (a) plate segment and (b) forces on plate segment

The quantity D is the bending stiffness of a plate. It reduces to the quantity EI, which is the bending stiffness of a beam of unit width, when we let $\mu = 0$.

7.5 Governing Differential Equations

The governing differential equation of a beam in bending is

$$\frac{d^2w}{dx^2} = -\frac{M(x)}{EI} \tag{7.19}$$

which can be expressed in terms of applied loads as

$$\frac{d^4w}{dx^4} = \frac{p(x)}{EI}. \tag{7.20}$$

A similar equation can be written for the bending of a plate. The corresponding differential equation for the bending of a plate is more complicated because it must include terms for the bending in the x- and y-directions as well as torsional moments that are present in the plate. Lagrange (Timoshenko 1983) was the first to develop the differential equation for the bending of a rectangular plate in 1811. We begin the derivation of the governing equations by considering an infinitesimal element dx, dy in Figure 7.7 subjected to lateral loads p. The forces and moments, per unit length, needed for equilibrium are shown in Figure 7.8 and are positive as shown. Also, downward deflection is taken as positive. It is of interest to note that two shearing forces, Q_x and Q_y, and two torsional moments, M_{xy} and M_{yx}, are needed to properly define the equilibrium of a rectangular plate.

Summation of forces in the z-direction gives the first equation of equilibrium:

$$p(x,y)dxdy - Q_x dy + \left(Q_x + \frac{\partial Q_x}{\partial x}dx \right) dy - Q_y dx + \left(Q_y + \frac{\partial Q_y}{\partial y}dy \right) dx = 0.$$

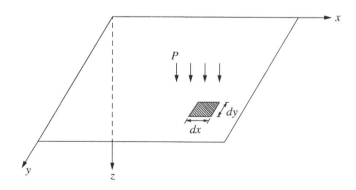

Figure 7.7 Lateral loads on a rectangular plate

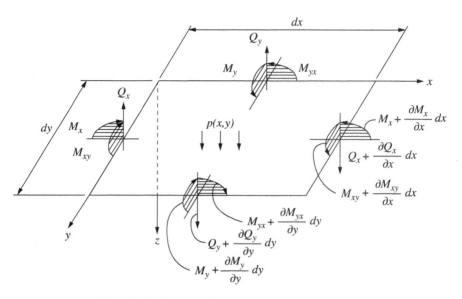

Figure 7.8 Forces and moments in a rectangular plate

This equation reduces to

$$p(x,y) + \frac{\partial Q_x}{\partial x} + \frac{\partial Q_y}{\partial y} = 0. \tag{7.21}$$

Summation of moments around x-axis gives the second equation of equilibrium:

$$M_y dx - \left(M_y + \frac{\partial M_y}{\partial y} dy\right) dx - M_{xy} dy + \left(M_{xy} + \frac{\partial M_{xy}}{\partial x} dx\right) dy$$

$$+ \left(Q_y + \frac{\partial Q_y}{\partial y} dy\right) dx\, dy - Q_x dy\, dy/2$$

$$+ \left(Q_x + \frac{\partial Q_x}{\partial x} dx\right) dy\, dy/2 + p dx\, dy\, dy/2 = 0.$$

Simplifying this equation gives

$$Q_y + \frac{\partial M_{xy}}{\partial x} - \frac{\partial M_y}{\partial y} + \left(\frac{\partial Q_y}{\partial y} + \frac{1}{2}\frac{\partial Q_x}{\partial x} + \frac{1}{2}p\right) dy = 0.$$

The bracketed term in this equation is multiplied by an infinitesimal quantity dy. It can thus be deleted because its magnitude is substantially less than that of the other three terms. The equation becomes

$$\frac{\partial Q_y}{\partial y} = \frac{\partial^2 M_y}{\partial y^2} - \frac{\partial^2 M_{xy}}{\partial x\,\partial y}. \tag{7.22}$$

Summation of moments around the y-axis gives the third equation of equilibrium:

$$\frac{\partial Q_x}{\partial x} = \frac{\partial^2 M_x}{\partial x^2} - \frac{\partial^2 M_{yx}}{\partial x\,\partial y}. \tag{7.23}$$

Substituting Eqs. (7.22) and (7.23) into Eq. (7.21) gives

$$p(x,y) + \frac{\partial^2 M_x}{\partial x^2} - 2\frac{\partial^2 M_{xy}}{\partial x\,\partial y} + \frac{\partial^2 M_y}{\partial y^2} = 0. \tag{7.24}$$

In this equation it is assumed that $M_{xy} = M_{yx}$ because at any point on the plate the shearing stress $\tau_{xy} = -\tau_{yx}$.

Substituting Eq. (7.17) into this equation gives

$$\frac{\partial^4 w}{\partial x^4} + 2\frac{\partial^4 w}{\partial x^2\,\partial y^2} + \frac{\partial^4 w}{\partial y^4} = \frac{p(x,y)}{D}. \tag{7.25}$$

A comparison of this equation with Eq. (7.20) for the bending of beams indicates that Eq. (7.25) is considerably more complicated because it considers the deflection in the x- and y-directions as well as the shearing effects in the xy plane.

Equation (7.25) can also be written as

$$\nabla^2 \nabla^2 w = \nabla^4 w = \frac{p(x,y)}{D} \tag{7.26}$$

where

$$\nabla^2 w = \frac{\partial^2 w}{\partial x^2} + \frac{\partial^2 w}{\partial y^2}$$

and

$$\nabla^4 w = \frac{\partial^4 w}{\partial x^4} + 2\frac{\partial^4 w}{\partial x^2 \partial y^2} + \frac{\partial^4 w}{\partial y^4}.$$

Equation (7.26) is the basic differential equation for rectangular plates in bending. A solution of this equation yields an expression for the deflection, w, of the plate. The moment expressions are obtained by substituting the deflection expressions into Eq. (7.17). The shear forces are obtained from Eqs. (7.22), (7.23), and (7.17) and are given by

$$Q_x = -D\left(\frac{\partial^3 w}{\partial x^3} + \frac{\partial^3 w}{\partial x \partial y^2}\right) \tag{7.27}$$

$$Q_y = -D\left(\frac{\partial^3 w}{\partial x^2 \partial y} + \frac{\partial^3 w}{\partial y^3}\right). \tag{7.28}$$

For sign convention we will assume a downward deflection as positive in Eq. (7.26). All other quantities are assumed positive as shown in Figure 7.8.

7.6 Boundary Conditions

The most frequently encountered boundary conditions for rectangular plates are essentially the same as those for beams. They are either fixed, simply supported, free, or partially fixed as shown in Figure 7.9.

a. Fixed edges: For a fixed edge (Figure 7.9), the deflection and slope are zero. Thus,

$$w\big|_{y=b} = 0 \tag{7.29}$$

$$\frac{\partial w}{\partial y}\big|_{y=b} = 0. \tag{7.30}$$

b. Simply supported edge: For a simply supported edge (Figure 7.9), the deflection and moment are zero. Hence,

$$w\big|_{y=0} = 0 \tag{7.31}$$

and, from Eq. (7.17),

$$M_y\big|_{y=0} = -D\left(\frac{\partial^2 w}{\partial y^2} + \mu\frac{\partial^2 w}{\partial x^2}\right)\bigg|_{y=0} = 0. \tag{7.32}$$

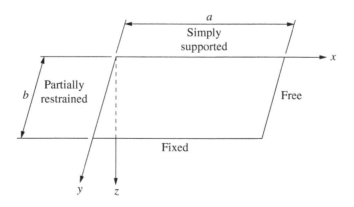

Figure 7.9   Plate with various boundary conditions

The expression $\mu \frac{\partial^2 w}{\partial x^2}$ in Eq. (7.32) can be written as $\mu \frac{\partial}{\partial x}\left(\frac{\partial w}{\partial x}\right)$, which is the rate of change of the slope at the boundary. But the change in slope along the simply supported edge $y = 0$ is always zero. Hence the quantity $\mu \frac{\partial^2 w}{\partial x^2}$ vanishes, and the moment boundary condition becomes

$$M_y\big|_{y=0} = \frac{\partial^2 w}{\partial y^2}\bigg|_{y=0} = 0. \tag{7.33}$$

c. Free edge: At a free edge, the moment and shear are zero. Hence,

$$M_x\big|_{x=a} = M_{xy}\big|_{x=a} = Q_x\big|_{x=a} = 0.$$

From the first of these boundary conditions and Eq. (7.17), we get

$$\left(\frac{\partial^2 w}{\partial x^2} + \mu \frac{\partial^2 w}{\partial y^2}\right)\bigg|_{x=a} = 0. \tag{7.34}$$

The other two boundary conditions can be combined into a single expression. Referring to Figure 7.10, it was shown by Kirchhoff (Timoshenko and Woinowsky-Krieger 1959) that the moment M_{xy} can be thought of as a series of couples acting on an infinitesimal section. Hence, at any point along the edge,

$$Q' = -\left(\frac{\partial M_{xy}}{\partial y}\right)\bigg|_{x=a}.$$

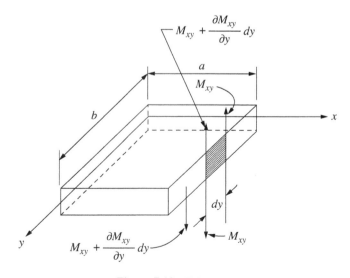

Figure 7.10 Edge shear

This equivalent shearing force, Q', must be added to the shearing force Q_x acting at the edge. Hence the total shearing force at the free edge is given by Q' and Eq. (7.27) as

$$V_x = \left(Q_x - \frac{\partial M_{xy}}{\partial y}\right)\Bigg|_{x=a} = 0.$$

Substituting the values of Q_x and M_{xy} from Eqs. (7.27) and (7.17) into this equation gives

$$\left(\frac{\partial^3 w}{\partial x^3} + (2-\mu)\frac{\partial^3 w}{\partial x\,\partial y^2}\right)\Bigg|_{x=a} = 0. \tag{7.35}$$

Equations (7.34) and (7.35) are the two necessary boundary conditions at a free edge of a rectangular plate.

d. Partially fixed edge: A partially fixed edge occurs in continuous plates or plates connected to beams. For this latter condition, Figure 7.11 shows that the two boundary conditions are given by

$$V\big|_{\text{plate}} = V\big|_{\text{beam}}$$

$$D\left[\frac{\partial^3 w}{\partial x^3} + (2-\mu)\frac{\partial^3 w}{\partial x\,\partial y^2}\right]\Bigg|_{x=0} = EI\left(\frac{\partial^4 w}{\partial y^4}\right)\Bigg|_{x=0}. \tag{7.36}$$

and

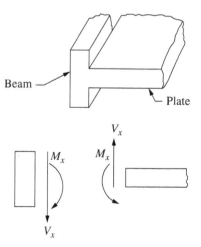

Figure 7.11 Plate with edge beam support

$$M\big|_{\text{plate}} = M\big|_{\text{beam}}$$

$$D\left(\frac{\partial^2 w}{\partial x^2} + \mu\frac{\partial^2 w}{\partial y^2}\right)\Bigg|_{x=0} = GJ\left(\frac{\partial^3 w}{\partial x\,\partial y^2}\right)\Bigg|_{x=0}. \tag{7.37}$$

e. Corner reactions: It was shown in the derivation of Eq. (7.35) that the torsion moment M_{xy} shown in Figure 7.10 can be resolved into a series of couples. At any corner, say $x = a$ and $y = b$ in Figure 7.12, the moment M_{xy} results in a downward force and so does M_{yx} as shown in the figure. Hence the total reaction at $x = a$ and $y = b$ is given by

$$R = 2\left(M_{xy}\right)\Big|_{\substack{x=a\\y=b}} = 2D(1-\mu)\left(\frac{\partial^2 w}{\partial x\,\partial y}\right)\Big|_{\substack{x=a\\y=b}}. \tag{7.38}$$

Equation (7.38) is normally used to determine the force in corner bolts of rectangular cover plates of gear transmission casings, flanges, etc.

The procedure detailed previously for determining the stress in a rectangular plate can be summarized as follows:

1. Express the deflection, w, and loads, p, in Eq. (7.26) in terms of a Fourier series.
2. Solve Eq. (7.26) for deflection, w, using the following boundary conditions:
   a. For a fixed edge: Eqs. (7.29) and (7.30)
   b. For a simply supported edge: Eqs. (7.31) and (7.32)
   c. For a free edge: Eqs. (7.34) and (7.35)
   d. For a plate attached to an edge beam: Eqs. (7.36) and (7.37)
3. Calculate the bending moment expressions from Eqs. (7.17).
4. Calculate the bending stress from the expression $S = 6\,M/t^2$ where M is the bending moment and t is the plate thickness.

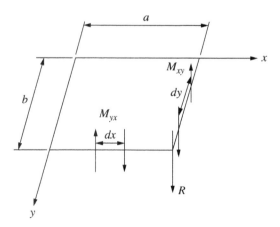

Figure 7.12   Edge shear moments

7.7 Double Series Solution of Simply Supported Plates

The first successful solution of a simply supported rectangular plate subjected to uniform load was made by Navier (Timoshenko 1983) in 1820. He assumed the load p in Eq. (7.26) to be represented by the double Fourier series, Appendix A, of the form

$$p(x,y) = \sum_{m=1}^{\infty} \sum_{n=1}^{\infty} p_{mn} \sin\frac{m\pi x}{a} \sin\frac{n\pi y}{b} \tag{7.39}$$

where p_{mn} is obtained from

$$p_{mn} = \frac{4}{ab} \int_0^b \int_0^a f(x,y) \sin\frac{m\pi x}{a} \sin\frac{n\pi y}{b} dxdy \tag{7.40}$$

and $f(x, y)$ is the shape of the applied load.
  Similarly the deflection w is expressed by

$$w(x,y) = \sum_{m=1}^{\infty} \sum_{n=1}^{\infty} w_{mn} \sin\frac{m\pi x}{a} \sin\frac{n\pi y}{b}. \tag{7.41}$$

  This equation automatically satisfies four boundary conditions of a simply supported plate, and w_{mn} is a constant that is determined from the differential equation.
  The solution of a rectangular plate problem consists of obtaining a load function from Eq. (7.39). The unknown constant w_{mn} is obtained by substituting Eqs. (7.39) and (7.41) into Eq. (7.26).

Example 7.1
a. Determine the maximum bending moment of a simply supported plate due to a uniformly applied load.
b. Let a steel rectangular plate with dimensions $a = 30$ inch and $b = 20$ inch be subjected to a pressure of 15 psi. Determine the maximum bending moment and deflection if $\mu = 0.3$, $E = 30,000$ ksi, and $t = 0.38$ inch.

Solution
a. Let the coordinate system be as shown in Figure 7.13. Equation (7.40) can be solved by letting $f(x, y)$ equal a constant p_o because the load is uniform over the entire plate. Hence,

$$p_{mn} = \frac{4p_o}{ab} \int_0^b \int_0^a \sin\frac{m\pi x}{a} \sin\frac{n\pi y}{b} dxdy = \frac{4p_o}{\pi^2 mn}(\cos m\pi - 1)(\cos n\pi - 1) = \frac{16p_o}{\pi^2 mn}$$

where
$m = 1, 3, 5,\ldots$
$n = 1, 3, 5,\ldots$

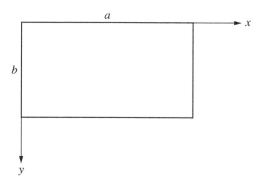

Figure 7.13   Uniformly loaded plate

From Eq. (7.39),

$$p = \frac{16p_\text{o}}{\pi^2} \sum_{m=1,3,\dots}^{\infty} \sum_{n=1,3,\dots}^{\infty} \frac{1}{mn} \sin\frac{m\pi x}{a} \sin\frac{n\pi y}{b}.$$

Substituting this equation and Eq. (7.41) into Eq. (7.26) gives

$$w_{mn} = \frac{16p_\text{o}}{\pi^6 mn D\left[(m/a)^2 + (n/b)^2\right]^2} \quad \begin{array}{l} m = 1,3,5,\dots \\ n = 1,3,5,\dots \end{array}$$

Hence, the deflection expression becomes

$$w = \frac{16p_\text{o}}{\pi^6 D} \sum_{m=1,3,\dots}^{\infty} \sum_{n=1,3,\dots}^{\infty} \frac{\sin\frac{m\pi x}{a} \sin\frac{n\pi y}{b}}{mn\left[(m/a)^2 + (n/b)^2\right]^2}. \tag{1}$$

The bending and torsional moment expressions are given by Eq. (7.17) and are expressed as

$$M_x = \frac{16p_\text{o}}{\pi^4} \left[\sum_{m=1,3,\dots}^{\infty} \sum_{n=1,3,\dots}^{\infty} F_{mn} \sin\frac{m\pi x}{a} \sin\frac{n\pi y}{b} \right] \tag{2}$$

$$M_y = \frac{16p_\text{o}}{\pi^4} \left[\sum_{m=1,3,\dots}^{\infty} \sum_{n=1,3,\dots}^{\infty} G_{mn} \sin\frac{m\pi x}{a} \sin\frac{n\pi y}{b} \right] \tag{3}$$

$$M_{xy} = \frac{16p_\text{o}(1-\mu)}{\pi^4} \left[\sum_{m=1,3,\dots}^{\infty} \sum_{n=1,3,\dots}^{\infty} H_{mn} \cos\frac{m\pi x}{a} \cos\frac{n\pi y}{b} \right] \tag{4}$$

where

$$F_{mn} = \frac{(m/a)^2 + \mu(n/b)^2}{mn\left[(m/a)^2 + (n/b)^2\right]^2}$$

$$G_{mn} = \frac{\mu(m/a)^2 + (n/b)^2}{mn\left[(m/a)^2 + (n/b)^2\right]^2}$$

$$H_{mn} = \frac{1}{ab\left[(m/a)^2 + (n/b)^2\right]^2}.$$

b. The maximum values of M_x, M_y, and w are obtained from expressions (2), (3), and (4) as

$$M_x = \frac{16p_o b^2}{\pi^4}(0.3035) = 299.1 \text{ inch-lbs/inch}$$

$$M_y = \frac{16p_o b^2}{\pi^4}(0.4941) = 487.0 \text{ inch-lbs/inch}$$

$$w = \frac{16p_o b^4}{\pi^6 D}(0.4647).$$

The value of D is given by Eq. (7.18) as

$$D = \frac{Et^3}{12(1-\mu^2)} = \frac{30,000,000 \times 0.38^3}{12(0.91)} = 150,747 \text{ lbs-inch.}$$

The maximum deflection is

$$w = 0.12 \text{ inch.}$$

The maximum values of the deflection and bending moments in a simply supported plate occur in the middle, Figure 7.13, where $x = a/2$ and $y = b/2$. Table 7.2 lists the coefficients for maximum deflection and maximum stress of a uniformly loaded simply supported plate.

It is of interest to note in Table 7.2 that for $a/b = 4$, the stress coefficient $K_1 = 0.741$ is within 1% of the coefficient 0.75 obtained for a uniformly loaded simply supported beam having the same length as the short edge of the plate.

7.8 Single Series Solution of Simply Supported Plates

Levy (Timoshenko 1983) in 1900 developed a method for solving simply supported plates subjected to various loading conditions using single Fourier series. This method is more practical then Navier's solution because it can also be used in plates with various boundary conditions. Levy suggested the solution of Eq. (7.26) to be expressed in terms of homogeneous and

Table 7.2 Maximum stress and deflection coefficients in a simply supported plate ($\mu = 0.3$)

a/b	$S = K_1 p_o b^2 / t^2$ K_1	$w = K_2 p_o b^4 / E t^3$ K_2
1.0	0.287	0.0444
1.1	0.329	0.0532
1.2	0.376	0.0617
1.3	0.416	0.0698
1.4	0.453	0.0774
1.5	0.487	0.0844
1.6	0.517	0.0907
1.8	0.569	0.102
2.0	0.610	0.111
2.5	0.678	0.126
3.0	0.713	0.134
3.5	0.731	0.138
4.0	0.741	0.140

a, long edge of plate; b, short edge of plate; E, modulus of elasticity; P_o, uniform load, S, maximum stress; t, thickness; w, maximum deflection.

particular parts each of which consists of a single Fourier series where the unknown function is determined from the boundary conditions. The solution is expressed as

$$w = w_h + w_p. \tag{7.42}$$

The homogeneous solution is written as

$$w_h = \sum_{m=1}^{\infty} f_m(y) \sin \frac{m\pi x}{a} \tag{7.43}$$

where $f(y)$ indicates that it is a function of y only. This equation also satisfies a simply supported boundary condition at $x = 0$ and $x = a$. Substituting Eq. (7.43) into the differential equation

$$\nabla^4 w = 0$$

gives

$$\left[\left(\frac{m\pi}{a}\right)^4 f_m(y) - 2\left(\frac{m\pi}{a}\right)^2 \frac{d^2 f_m(y)}{dy^2} + \frac{d^4 f_m(y)}{dy^4} \right] \sin \frac{m\pi x}{a} = 0$$

which is satisfied when the bracketed term is equal to zero. Thus,

$$\frac{d^4 f_m(y)}{dy^4} - 2\left(\frac{m\pi}{a}\right)^2 \frac{d^2 f_m(y)}{dy^2} + \left(\frac{m\pi}{a}\right)^4 f_m(y) = 0. \tag{7.44}$$

The solution of this differential equation can be expressed as

$$f_m(y) = F_m e^{R_m y}.$$

(7.45)

Substituting Eq. (7.45) into Eq. (7.44) gives

$$R_m^4 - 2\left(\frac{m\pi}{a}\right)^2 R_m^2 + \left(\frac{m\pi}{a}\right)^4 = 0$$

which has the roots

$$R_m = \pm\frac{m\pi}{a}, \quad \pm\frac{m\pi}{a}.$$

Thus, the solution of Eq. (7.44) is

$$f_m(y) = C_{1m} e^{\frac{m\pi y}{a}} + C_{2m} e^{-\frac{m\pi y}{a}} + C_{3m} y e^{\frac{m\pi y}{a}} + C_{4m} y e^{-\frac{m\pi y}{a}}$$

where C_{1m}, C_{2m}, C_{3m}, and C_{4m} are constants. This equation can also be written as

$$f_m(y) = A_m \sinh\frac{m\pi y}{a} + B_m \cosh\frac{m\pi y}{a} + C_m y \sinh\frac{m\pi y}{a} + D_m y \cosh\frac{m\pi y}{a}.$$

The homogeneous solution given by Eq. (7.43) becomes

$$w_h = \sum_{m=1}^{\infty} \left[A_m \sinh\frac{m\pi y}{a} + B_m \cosh\frac{m\pi y}{a} + C_m y \sinh\frac{m\pi y}{a} + D_m y \cosh\frac{m\pi y}{a} \right] \sin\frac{m\pi x}{a}$$

(7.46)

where the constants A_m, B_m, C_m, and D_m are obtained from the boundary conditions.

The particular solution, w_p, in Eq. (7.42) can be expressed by a single Fourier series as

$$w_p = \sum_{m=1}^{\infty} k_m(y) \sin\frac{m\pi x}{a}.$$

(7.47)

The load p is expressed as

$$p(x,y) = \sum_{m=1}^{\infty} p_m(y) \sin\frac{m\pi x}{a}$$

(7.48)

where

$$p_m(y) = \frac{2}{a} \int_0^a p(x,y) \sin\frac{m\pi x}{a} dx.$$

(7.49)

Substituting Eqs. (7.47) and (7.48) into Eq. (7.26) gives

$$\frac{d^4 k_m}{dy^4} - 2\left(\frac{m\pi}{a}\right)^2 \frac{d^2 k_m}{dy^2} + \left(\frac{m\pi}{a}\right)^4 k_m = \frac{p_m(y)}{D}. \tag{7.50}$$

Thus, the solution of the differential equation (7.26) consists of solving Eqs. (7.46) and (7.50) as shown in the following example.

Example 7.2
The rectangular plate shown in Figure 7.14 is subjected to a uniform load p_o. Determine the expression for the deflection.

Solution
From Eq. (7.49),

$$p_m(y) = \frac{2p_o}{a} \int_0^a \sin\frac{m\pi x}{a} dx = \frac{2p_o}{m\pi}(\cos m\pi - 1) = \frac{4p_o}{m\pi} \quad m = 1, 3, \ldots$$

Hence, Eq. (7.50) becomes

$$\frac{d^4 k_m}{dy^4} - 2\left(\frac{m\pi}{a}\right)^2 \frac{d^2 k_m}{dy^2} + \left(\frac{m\pi}{a}\right)^4 k_m = \frac{4p_o}{m\pi D}. \tag{1}$$

The particular solution of this equation can be taken as

$$k_m = C.$$

Substituting this expression into Eq. (1) gives

$$k_m = \frac{4a^4 p_o}{m^5 \pi^5 D} \quad m = 1, 3, \ldots$$

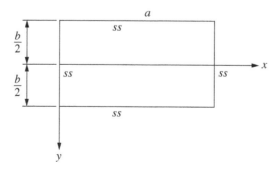

Figure 7.14 Uniformly loaded simply supported plate

And Eq. (7.47) for the particular solution becomes

$$w_p = \frac{4a^4 p_o}{\pi^5 D} \sum_{m=1,3,\ldots}^{\infty} \frac{1}{m^5} \sin\frac{m\pi x}{a}. \tag{2}$$

The homogeneous solution for the deflection is obtained from Eq. (7.46). Referring to Figure 7.14, the deflection in the y-direction due to uniform load is symmetric about the x-axis. Hence, the constants A_m and D_m must be set to zero since the quantities $\sinh\frac{m\pi y}{a}$ and $y\cosh\frac{m\pi y}{a}$ are odd functions as y varies from positive to negative. Also, m must be set to 1, 3, 5,… in order for $\sin\frac{m\pi x}{a}$ to be symmetric around $x = a/2$. Hence,

$$w_h = \sum_{m=1,3,\ldots}^{\infty} \left(B_m \cosh\frac{m\pi y}{a} + C_m y\sinh\frac{m\pi y}{a} \right) \sin\frac{m\pi x}{a}$$

and the total deflection can now be expressed as

$$w = \sum_{m=1,3,\ldots}^{\infty} \left(B_m \cosh\frac{m\pi y}{a} + C_m \sinh\frac{m\pi y}{a} + \frac{4p_o a^4}{m^5 \pi^5 D} \right) \sin\frac{m\pi x}{a}. \tag{3}$$

The boundary conditions along the y-axis are expressed as

$$w = 0 \quad \text{at} \quad y = \pm b/2$$

and

$$\frac{\partial^2 w}{\partial y^2} = 0 \quad \text{at} \quad y = \pm b/2.$$

From the first of these boundary conditions, we get

$$B_m \cosh\frac{m\pi b}{2a} + C_m \frac{b}{2}\sinh\frac{m\pi b}{2a} + \frac{4a^4 p_o}{m^5 \pi^5 D} = 0$$

and from the second boundary condition, we obtain

$$\left[B_m \left(\frac{m\pi}{a} \right) + bC_m \right] \cosh\frac{m\pi b}{2a} + C_m \left(\frac{m\pi b}{2a} \right) \sinh\frac{m\pi b}{2a} = 0.$$

Solving these two simultaneous equations yields

$$C_m = \frac{2a^3 p_o}{m^4 \pi^4 D \cosh(m\pi b/2a)}$$

and

$$B_m = \frac{4a^4 p_o + m\pi p_o a^3 b \tanh(m\pi b/2a)}{m^5 \pi^5 D \cosh(m\pi b/2a)}.$$

With these two expressions known, Eq. (3) can now be solved for various values x and y.

7.9   Rectangular Plates with Fixed Edges

The deflection of a rectangular plate uniformly loaded with fixed edges is obtained by super-imposing three loading conditions (Timoshenko and Woinowsky-Krieger 1959) using the Levy solution. The deflection of the first loading condition is that of a simply supported plate uniformly loaded as given in Example 7.2. The deflection of the second loading condition is that of a simply supported plate with bending moments, M_1, applied on the two short edges. M_1 is expressed in a Fourier series with an unknown coefficient M_{1m}. The deflection of the third loading condition is that of a simply supported plate with bending moments, M_2, applied on the two long edges. M_2 is expressed in a Fourier series with an unknown coefficient M_{2n}. The magnitude of the moment coefficients M_{1m} and M_{2n} is determined by setting the rotation at the fixed edges due to all three loading conditions to zero. The solution is tedious but straightforward. The resulting maximum deflection at the middle of the plate and the maximum bending moment at the middle of the short edges are shown in Table 7.3.

When $a/b = 2$ in Table 7.2, the stress coefficient $K_3 = 0.497$ is within 1% of the coefficient 0.5 for a uniformly loaded beam with fixed ends and a length equal to the short edge of the plate.

Table 7.3   Maximum stress and deflection coefficients in a fixed plate ($\mu = 0.3$)

a/b	$S = K_3 p_o b^2/t^2$ K_3	$w = K_4 p_o b^4/Et^3$ K_4
1.0	0.308	0.0138
1.1	0.380	0.0168
1.2	0.383	0.0188
1.3	0.409	0.0206
1.4	0.436	0.0226
1.5	0.454	0.0238
1.6	0.468	0.0251
1.8	0.487	0.0267
2.0	0.497	0.0277
2.5	0.500	0.0279
3.0	0.500	0.0281
3.5	0.500	0.0283
4.0	0.500	0.0284

a, long edge of plate; b, short edge of plate; E, modulus of elasticity; P_o, uniform load; S, maximum stress; t, thickness; w, maximum deflection.

7.10   Plate Equations in the ASME Code

7.10.1   Boiler Code: Section I

Section I lists the following equation in paragraph PG-31.3.3 for designing rectangular plates:

$$t = d \left(\frac{ZCP}{S} \right)^{0.5} \tag{7.51}$$

where
C = coefficient that is a function of rectangular head-to-shell junction and varies between 0.17
      and 0.33
d = short span
D = large span
P = internal pressure
S = allowable stress
t = thickness
$Z = 3.4 - 2.4d/D$

For the case $D/d = 2.0$, the expression for stress from Eq. (7.51) for $C = 0.33$ (Figure PG-31(i-2) of Section I) is $S = 0.73Pd^2/t^2$. The expression for a simply supported rectangular plate from Table 7.2 is $S = 0.61Pd^2/t^2$, while that for a fixed rectangular plate from Table 7.3 is $S = 0.50Pd^2/t^2$. Similarly the expression for a simply supported beam is $S = 0.75Pd^2/t^2$, while that for a fixed beam is $S = 0.5Pd^2/t^2$. The coefficients for these various cases are shown in Figure 7.15.

7.10.2   Nuclear Code: Section III

7.10.2.1   Subsections NB and NC

Rules are not given for a rectangular plate.

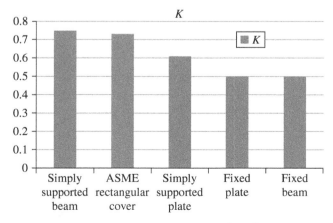

Figure 7.15   Coefficient K for beams and for the equation $S = d^2KP/t^2$ for uniformly loaded rectangular plates with $D/d = 2.0$.

7.10.2.2 Subsection ND

The rules are essentially the same as those in Section I.

7.10.3 Pressure Vessel Code: Section VIII

7.10.3.1 Divisions 1 and 2

The rules are essentially the same as those in Section I.

Example 7.3

What is the required cover thickness of a rectangular vessel with $D = 24$ inch, $d = 18$ inch, $P = 250$ psi, and $S = 20,000$ psi using (a) plate theory and (b) ASME rules with $C = 0.33$?

Solution

a. $a = 24$ inch, $b = 18$ inch, $a/b = 1.3$

Assuming a conservative boundary condition of simply supported plate. Then from Table 7.1, $K_1 = 0.416$.

$$t = \left[(0.416)(250)(18^2)/20,000\right]^{0.5}$$

$$= 1.30 \text{ inch.}$$

b. From Eq. (7.51) with $d = 18$ inch and $D = 24$ inch,

$$Z = 3.4 - 2.4\left(\frac{18}{24}\right)$$

$$= 1.6.$$

$$t = (18)[(1.6)(0.33)(250)/20,000]^{0.5}$$

$$= 1.46 \text{ inch.}$$

Problems

7.1 The finite element formulation for the stiffness of a solid three-dimensional element is based on the strain–stress matrix Eq. (7.13). Rewrite this equation as a stress–strain matrix.

7.2 Determine the maximum bending stress values σ_x and σ_y in a simply supported plate, with length $a = 100$ cm and width $b = 75$ cm. The deflection is approximated by

$$w = k \sin\frac{\pi x}{a} \sin\frac{\pi y}{b}$$

where k is a constant equal to 0.462 cm. Let $t = 1.2$ cm, $\mu = 0.3$, and $E = 200,000$ MPa.

7.3 Find M_x, M_y, M_{xy}, Q_x, and Q_y of a rectangular plate whose deflection is given by

$$w = k \sin\frac{m\pi x}{a}\sin\frac{n\pi y}{b}$$

where k, a, b, n, and m are constants.

7.4 Derive Eqs. (7.25), (7.27), and (7.28).

7.5 Find the expression for M_x, M_y, and M_{xy} in Example 7.2.

8

Bending of Circular Plates

8.1 Plates Subjected to Uniform Loads in the θ-Direction

Circular plates are common in many structures such as nozzle covers, end closures in pressure vessels, and bulkheads in submarines and airplanes. The derivation of the classical equations for lateral bending of circular plates dates back to 1828 and is accredited to Poisson (Timoshenko 1983). He used polar coordinates to transfer the differential equations for the bending of a rectangular plate to circular plates. The first rigorous solution of the differential equation of circular plates for various loading and boundary conditions was made around 1900 and is credited to Love (1944).

The basic assumptions made in deriving the differential equations for lateral bending of rectangular plates in Section 7.1 are also applicable to circular plates. The differential equations for the lateral bending of circular plates subjected to uniform loads in the θ-direction are derived from Figure 8.1. For sign convention it will be assumed that downward deflections and clockwise rotations are positive. Hence, if a flat plate undergoes a small deflection as shown in Figure 8.1, then the radius of curvature r at point B is given by

$$\sin(\phi) \approx \phi = \frac{r}{r_\theta}$$

or

$$\frac{1}{r_\theta} = \frac{\phi}{r} = \frac{-1}{r}\frac{dw}{dr}. \tag{8.1}$$

The quantity r_θ represents a radius that forms a cone as it rotates around the z-axis (in and out of the plane of the paper). The second radius of curvature is denoted by r_r. The origin of r_r does

Stress in ASME Pressure Vessels, Boilers, and Nuclear Components, First Edition. Maan H. Jawad.
© 2018, The American Society of Mechanical Engineers (ASME), 2 Park Avenue,
New York, NY, 10016, USA (www.asme.org). Published 2018 by John Wiley & Sons, Inc.

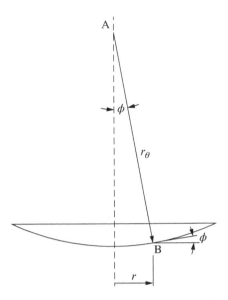

Figure 8.1 Lateral bending of a circular plate

not necessarily fall on the axis of symmetry although, for any point B, the radii r_r and r_θ coincide with each other. The value of r_r is obtained from Eq. (7.6) as

$$\chi = \frac{1}{r_r} = -\frac{d^2w}{dr^2}$$

or

$$\chi = \frac{1}{r_r} = -\frac{d^2w}{dr^2} = \frac{d\phi}{dr}. \tag{8.2}$$

The M_x and M_y expressions in Eq. (7.17) can be written in terms of the radial and tangential directions as

$$\begin{bmatrix} M_r \\ M_t \end{bmatrix} = D \begin{bmatrix} 1 & \mu \\ \mu & 1 \end{bmatrix} \begin{bmatrix} \dfrac{1}{r_r} \\ \dfrac{1}{r_\theta} \end{bmatrix} \tag{8.3}$$

or

$$\begin{bmatrix} M_r \\ M_t \end{bmatrix} = D \begin{bmatrix} 1 & \dfrac{\mu}{r} \\ \mu & \dfrac{1}{r} \end{bmatrix} \begin{bmatrix} \dfrac{d\phi}{dr} \\ \phi \end{bmatrix} \tag{8.4}$$

and

$$\begin{bmatrix} M_r \\ M_t \end{bmatrix} = -D \begin{bmatrix} 1 & \dfrac{\mu}{r} \\ \mu & \dfrac{1}{r} \end{bmatrix} \begin{bmatrix} \dfrac{d^2w}{dr^2} \\ \dfrac{dw}{dr} \end{bmatrix} \tag{8.5}$$

where

$$D = \frac{Et^3}{12(1-\mu^2)}.$$

The classical theory of the lateral bending of circular plates discussed in this section is based on the assumption that the loads on the plate are uniformly distributed in the θ-direction. In this case the torsional moment $M_{r\theta}$ is zero, and the other forces are as shown in Figure 8.2a. Summing moments around line a–a gives

$$(M_r r d\theta) - \left(M_r + \frac{dM_r}{dr} dr\right)(r+dr)d\theta + 2(M_t \ dr \ d\theta/2)$$

(8.6)

$$-Qr \ d\theta \ dr/2 - \left(Q + \frac{dQ}{dr} dr\right)(r+dr)d\theta \ dr/2 = 0.$$

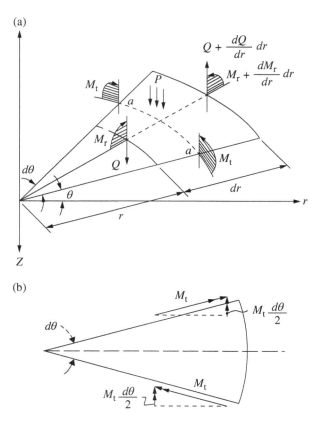

Figure 8.2 Forces and moments in a circular plate symmetrically loaded in the θ-direction: (a) infinitesimal element and (b) components of M_t

The quantity $M_t\, dr\, d\theta/2$ is the component of M_t perpendicular to axis a–a as shown in Figure 8.2b. Equation (8.6) can be reduced to

$$M_r + \frac{dM_r}{dr}r - M_t + Qr = 0. \tag{8.7}$$

Substituting Eq. (8.5) into this equation gives

$$\frac{d}{dr}\left[\frac{1}{r}\frac{d}{dr}\left(r\frac{dw}{dr}\right)\right] = \frac{Q}{D}. \tag{8.8}$$

Or since

$$2\pi r Q = \int p 2\pi r\, dr,$$

Equation (8.8) can be written in a different form as

$$\frac{1}{r}\frac{d}{dr}\left\{r\frac{d}{dr}\left[\frac{1}{r}\frac{d}{dr}\left(r\frac{dw}{dr}\right)\right]\right\} = \frac{p}{D} \tag{8.9}$$

where p is a function of r.

The analysis of circular plates with uniform thickness subjected to symmetric lateral loads consists of solving the differential equation for the deflection as given by Eq. (8.8) or (8.9). The bending moments are then calculated from Eq. (8.5). The shearing force is calculated from Eqs. (8.7) and (8.5) as

$$Q = D\left(\frac{d^3w}{dr^3} + \frac{1}{r}\frac{d^2w}{dr^2} - \frac{1}{r^2}\frac{dw}{dr}\right) \tag{8.10}$$

or, from Eqs. (8.7) and (8.4), as

$$Q = -D\left(\frac{d^2\phi}{dr^2} + \frac{1}{r}\frac{d\phi}{dr} - \frac{\phi}{r^2}\right). \tag{8.11}$$

Example 8.1
a. Find the expression for the maximum moment and deflection of a uniformly loaded circular plate with simply supported edges.
b. Find the required thickness of a steel plate if the allowable stress = 15,000 psi, $p = 5$ psi, $a = 20$ inches, $E = 30,000,000$ psi, and $\mu = 0.3$. What is the maximum deflection?

Solution
a. From Figure 8.3, the shearing force Q at any radius r is given by

$$2\pi r Q = \pi r^2 p$$

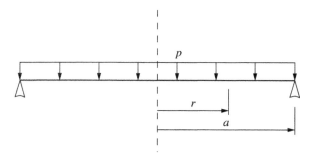

Figure 8.3   Uniformly loaded plate

or

$$Q = \frac{pr}{2}.$$

From Eq. (8.8)

$$\frac{d}{dr}\left[\frac{1}{r}\frac{d}{dr}\left(r\frac{dw}{dr}\right)\right] = \frac{pr}{2D}.$$

Integrating this equation gives

$$\text{Slope} = \frac{dw}{dr} = \frac{pr^3}{16D} + \frac{C_1 r}{2} + \frac{C_2}{r} \qquad (1)$$

$$\text{Deflection} = w = \frac{pr^4}{64D} + \frac{C_1 r^2}{4} + C_2 \ln r + C_3. \qquad (2)$$

At $r = 0$ the slope is equal to zero due to symmetry. Hence, from Eq. (1), C_2 must be set to zero. At $r = a$, the moment is zero and

$$-D\left(\frac{d^2 w}{dr^2} + \frac{\mu}{r}\frac{dw}{dr}\right) = 0$$

or

$$C_1 = \frac{-(3+\mu)pa^2}{(1+\mu)\ 8D}.$$

At $r = a$, the deflection is zero and Eq. (2) gives

$$C_3 = \frac{pa^4}{64D}\left(\frac{6+2\mu}{1+\mu} - 1\right).$$

The expression for deflection becomes

$$w = \frac{p}{64D}\left(a^2 - r^2\right)\left(\frac{5+\mu}{1+\mu}a^2 - r^2\right).$$

Substituting this expression into Eq. (8.5) gives

$$M_r = \frac{p}{16}(3+\mu)\left(a^2 - r^2\right)$$

$$M_t = \frac{p}{16}\left[a^2(3+\mu) - r^2(1+3\mu)\right].$$

The maximum deflection is at $r = 0$ and is given by

$$\max w = \frac{pa^4}{64D}\left(\frac{5+\mu}{1+\mu}\right).$$

A plot of M_r and M_t is shown in Figure 8.4 for $\mu = 0.3$. The plot shows that the maximum moment is in the center and is equal to

$$M_r = M_t = \frac{pa^2}{16}(3+\mu).$$

It is of interest to note that M_t is not zero at the edge of the plate. This is important in composite plates as reinforcing elements are needed around the perimeter to resist the tension stress caused by M_t.

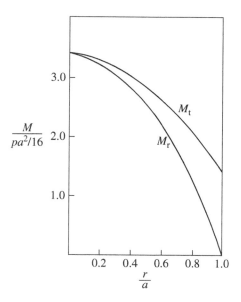

Figure 8.4 Plot of M_r and M_t of a uniformly loaded plate

b.

$$\text{Maximum } M = \frac{5 \times 20^2}{16}(3.3) = 412.5 \text{ inch-lbs/inch}$$

and

$$t = \sqrt{\frac{6M}{\sigma}}$$

$$t = \sqrt{\frac{6 \times 412.5}{15,000}}$$

$$= 0.41 \text{ inch}.$$

$$\text{Maximum } w = \frac{pa^4}{64D}\left(\frac{5+\mu}{1+\mu}\right) = 0.27 \text{ inch}.$$

Example 8.2

Find the stress at $r = a$ and $r = b$ for the plate shown in Figure 8.5. Let $a = 24$ inches, $b = 12$ inches, $F = 20$ lbs/inch, $t = 0.50$ inch, $E = 30,000$ ksi, and $\mu = 0.3$.

Solution

The shearing force at any point is given by

$$Q = \frac{bF}{r}.$$

Substituting this expression into Eq. (8.8) and integrating results in

$$w = \frac{bF}{4D}r^2(\ln r - 1) + \frac{C_1 r^2}{4} + C_2 \ln r + C_3. \tag{1}$$

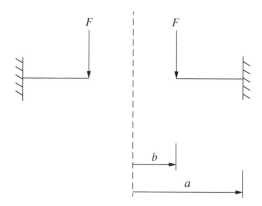

Figure 8.5   Plate loaded at its inner edge

The boundary conditions are

$$w = 0 \text{ at } r = a$$

$$\frac{dw}{dr} = 0 \text{ at } r = a$$

$$M_r = 0 \text{ at } r = b.$$

Evaluating Eq. (1) and its derivatives at the boundary conditions results in

$$C_1 = -648.53/D$$

$$C_2 = 1667.46/D$$

$$C_3 = 12,814.46/D.$$

At $r = b$,

$$M_r = 0$$

$$M_t = -D\left(\frac{1}{r}\frac{dw}{dr} + \mu\frac{d^2w}{dr^2}\right)$$

$$= -D\left\{\frac{bF}{2D}\ln r - \frac{bF}{2D}\frac{1}{2}\right.$$

$$\left. + \mu\left[\frac{bF}{2D}\left(\frac{1}{2} + \ln r\right) + \frac{C_1}{2} - \frac{C_2}{r^2}\right]\right\}$$

or

$$M_t = -67.8 \text{ inch-lbs/inch}$$

and

$$\sigma_t = \frac{6M}{t^2} = 1627 \text{ psi.}$$

Similarly, at $r = a$

$$M_t = -34.26 \text{ inch-lbs/inch}$$

$$\sigma_t = 822 \text{ psi}$$

$$M_r = -114.21 \text{ inch-lbs/inch}$$

$$\sigma_r = 2741 \text{ psi.}$$

Example 8.3
Find the expression for the deflection of the plate shown in Figure 8.6a due to load F.

Solution
The plate can be separated into two components (Figure 8.6b). Continuity between the inner and the outer plate is maintained by applying an unknown moment, M_o, as shown in

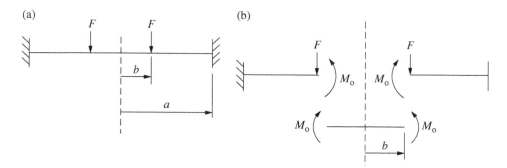

Figure 8.6 Plate loaded at distance b from the centerline: (a) plate with a line load and (b) free-body diagram

Figure 8.6b. The deflection of the inner plate due to M_o is obtained from Eq. (8.8) with $Q = 0$ as

$$w = \frac{C_1 r^2}{4} + C_2 \ln r + C_3. \tag{1}$$

The slope is

$$\frac{dw}{dr} = \frac{C_1 r}{2} + \frac{C_2}{r}, \tag{2}$$

and M_r is obtained from Eq. (8.5) as

$$M_r = -D\left[\frac{C_1}{2}(1+\mu) - \frac{C_2}{r^2}(1-\mu)\right]. \tag{3}$$

At $r = 0$, the slope is zero and from Eq. (2) we get $C_2 = 0$.
At $r = b$, $M_r = M_o$ and Eq. (3) yields

$$C_1 = \frac{-2M_o}{D(1+\mu)}. \tag{4}$$

Equations (1) and (2) can be expressed as

$$w = \frac{-M_o r^2}{2D(1+\mu)} + C_3 \tag{5}$$

$$\frac{dw}{dr} = \frac{-rM_o}{D(1+\mu)}. \tag{6}$$

The deflection of the outer plate is obtained from Example 8.2 as

$$w = \frac{bF}{4D}r^2(\ln r - 1) + \frac{C_4 r^2}{4} + C_5 \ln r + C_6. \tag{7}$$

At $r = a$, the slope is zero and Eq. (7) gives

$$\frac{C_4 a}{2} + \frac{C_5}{a} = \frac{-bFa}{4D}(2\ln a - 1). \tag{8}$$

At $r = b$, $M_r = M_o$ and from Eqs. (7) and (8.5) we get

$$\frac{-M_o}{D} = \frac{bF}{4D}[(1-\mu) + 2(1+\mu)\ln b] + \frac{C_4}{2}(1+\mu) + \frac{C_5}{b^2}(\mu - 1). \tag{9}$$

At $r = b$, the slope of the outer plate is equal to the slope of the inner plate. Taking the derivatives of Eq. (7) and equating it to Eq. (6) at $r = b$ gives

$$\frac{C_4 b}{2} + \frac{C_5}{b} = \frac{-b^2 F}{4D}(2\ln b - 1) - \frac{bM_o}{D(1+\mu)}. \tag{10}$$

Equations (8), (9), and (10) contain three unknowns. They are M_o, C_4, and C_5. Solving these three equations yields

$$M_o = \frac{(1+\mu)bF}{4a^2}\left[2a^2 \ln\left(\frac{a}{b}\right) - (a^2 - b^2)\right] \tag{11}$$

$$C_4 = \frac{bF}{D}\left(\frac{a^2 - b^2}{2a^2} - \ln a\right) \tag{12}$$

$$C_5 = \frac{b^3 F}{4D}. \tag{13}$$

With these quantities known, the other constants can readily be obtained. Constant C_1 is determined from Eq. (4). Constant C_6 is solved from Eq. (7) for the boundary conditions $w = 0$ at $r = a$. This gives

$$C_6 = \frac{bF}{8D}(a^2 + b^2 - 2b^2 \ln a). \tag{14}$$

Constant C_3 can now be calculated from the equation

$$w \text{ of inner plate}|_{r=b} = w \text{ of outer plate}|_{r=b}.$$

Equating Eqs. (1) and (7) at $r = b$ gives

$$C_3 = \frac{bF}{8a^2 D}\left[-2a^2 b^2 \ln\left(\frac{a}{b}\right) + (a^2 - b^2)a^2\right]. \tag{15}$$

Hence, the deflection of the inner plate is obtained by substituting Eqs. (11) and (15) into Eq. (5) to give

$$w = \frac{bF}{8a^2D}\left[\left(a^2 - b^2\right)\left(a^2 + r^2\right) - 2a^2\left(b^2 + r^2\right)\ln\left(\frac{a}{b}\right)\right],$$

and the deflection of the outer plate is obtained by substituting Eqs. (12), (13), and (14) into Eq. (7). This gives

$$w = \frac{bF}{8a^2D}\left[\left(a^2 + b^2\right)\left(a^2 - r^2\right) + 2a^2\left(b^2 + r^2\right)\ln\left(\frac{r}{a}\right)\right].$$

Table 8.1 shows a few loading conditions on circular plates.

8.2   Circular Plates in the ASME Code

The expression for the maximum stress in a simply supported plate is shown in Table 8.1. It can be written in terms of thickness as

$$t = d(0.3P/S)^{0.5} \tag{8.12}$$

where
d = diameter of plate
P = pressure
S = allowable stress
t = thickness

Similarly, the expression for the maximum stress in a plate with fixed edges is shown in Table 8.1. It can be written in terms of thickness as

$$t = d(0.19\ P/S)^{0.5}. \tag{8.13}$$

The ASME Section VIII-1 code shows various configurations (Perry 1950) of attachments of flat covers to cylindrical shells. The design equation is given by

$$t = d\left(\frac{CP}{SE}\right)^{0.5} \tag{8.14}$$

where

C = factor based on the attachment method of the flat plate to cylinder (The C factor in VIII-1 varies between 0.17 and 0.30 for most attachments. These values are within the range given by Eqs. (8.12) and (8.13) for simply supported and fixed plates)
E = joint efficiency of the weld in the flat plate

Table 8.1 Circular plates of uniform thickness

Case number	Maximum values
1. 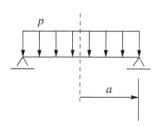	$S_r = S_t = \dfrac{3pa^2}{8t^2}(3+\mu)$ at center $w = \dfrac{3pa^4(1-\mu)(5+\mu)}{16Et^3}$ at center $\theta = -\dfrac{3pa^3(1-\mu)}{2Et^3}$ at edge For $\mu = 0.3$, $S_r = S_t = \dfrac{1.238pa^2}{t^2}$ at center $w = \dfrac{0.696pa^4}{Et^3}$ at center $\theta = -\dfrac{1.050pa^3}{Et^3}$ at edge
2. 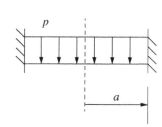	$S_r = -\dfrac{3pa^2}{4t^2}$ at edge $w = -\dfrac{3pa^4(1-\mu^2)}{16Et^3}$ at center For $\mu = 0.3$ $S_r = -\dfrac{3pa^2}{4t^2}$ at edge $w = \dfrac{0.171pa^4}{Et^3}$ at center
3. 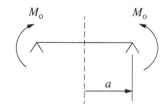	$S_r = S_t = 6M_o/t^2$ at any point $w = \dfrac{6M_o a^2(1-\mu)}{Et^3}$ at center $\theta = -\dfrac{12M_o a(1-\mu)}{Et^3}$ at edge For $\mu = 0.3$ $S_r = S_t = 6M_o/t^2$ at any point $w = \dfrac{4.20M_o a^2}{Et^3}$ at center $\theta = -\dfrac{8.40M_o a}{Et^3}$ at edge

Notation: θ, rotation, radians; $\mu = $ Poisson's ratio; a, outside radius of plate; E, modulus of elasticity; p, applied load; S_t, tangential stress; S_r, radial stress; t, thickness; and w, deflection.

8.3   Plates on an Elastic Foundation

Power and petrochemical plants as well as refineries use evaporators, condensers, and heat exchange units as part of their daily operations. These units consist of two perforated circular plates, called tubesheets, that are braced by a number of tubes as shown in Figure 8.7. The tubesheets and tubes are inserted in a vessel consisting of a cylindrical shell and two end closures. Fluid passing around the outside surface of the tubes exchanges heat with a different fluid passing through the tubes. The tubesheets are assumed to be supported by both the shell and tubes and are analyzed as circular plates on an elastic foundation. Referring to Figures 8.2 and 8.8, it is seen that the foundation pressure f acts opposite the applied pressure p. Hence Eq. (8.9) can be expressed as (Hetenyi 1964)

$$\frac{1}{r}\frac{d}{dr}\left\{r\frac{r}{dr}\left[\frac{1}{r}\frac{d}{dr}\left(r\frac{dw}{dr}\right)\right]\right\} = \frac{p_r - f_r}{D} \tag{8.15}$$

or

$$\left(\frac{d^2}{dr^2} + \frac{1}{r}\frac{d}{dr}\right)\left(\frac{d^2}{dr^2} + \frac{1}{r}\frac{d}{dr}\right)w = \frac{p_r - kw}{D} \tag{8.16}$$

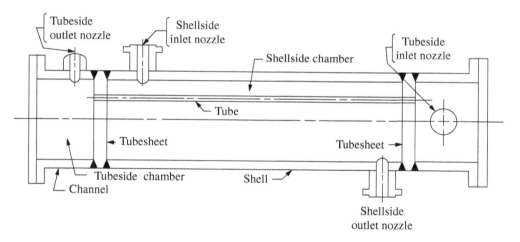

Figure 8.7   Heat exchanger

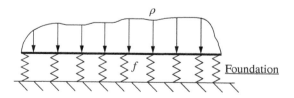

Figure 8.8   Plate on elastic foundation

where

f_r = load exerted by the elastic foundation

$f_r = k_o w$

k_o = foundation modulus defined as the modulus of elasticity of foundation divided by the depth of foundation, psi per inch

A plate on an elastic foundation that is subjected to uniform pressure will settle uniformly without developing any bending moments. If a support is placed at the edge of such a plate, then bending moments and shear are developed because of the nonuniform settlement caused by the boundary condition.

Accordingly, we can investigate the effects of the various boundary conditions on the plate stress by allowing the applied pressure to be set to zero. Letting

$$\alpha = \sqrt[4]{\frac{k_o}{D}}$$

the differential equation becomes

$$\left(\frac{d^2}{dr^2} + \frac{1}{r}\frac{d}{dr}\right)\left(\frac{d^2}{dr^2} + \frac{1}{r}\frac{d}{dr}\right)w + \alpha^4 w = 0.$$

Let $\alpha r = \sqrt[4]{-1}\rho$.

Then the differential equation becomes

$$\nabla^4 w - w = 0 \tag{8.17}$$

where

$$\nabla^4 = \left(\frac{d^2}{d\rho} + \frac{1}{\rho}\frac{d}{d\rho}\right)^2.$$

Equation (8.17) can be written either as

$$\nabla^2\left(\nabla^2 w + w\right) - \left(\nabla^2 w + w\right) = 0$$

or

$$\nabla^2\left(\nabla^2 w - w\right) + \left(\nabla^2 w - w\right) = 0.$$

The solution is a combination of

$$\nabla^2 w + w = 0$$

and

$$\nabla^2 w - w = 0.$$

The first equation can be written as

$$\frac{d^2w}{d\rho^2} + \frac{1}{\rho}\frac{dw}{d\rho} + w = 0,$$

and the solution is expressed in terms of Bessel function (see Appendix B) as

$$w = A_1 J_o(\rho) + A_2 Y_o(\rho).$$

The second equation has a solution in the form of

$$w = A_3 J_o(i\rho) + A_4 Y_o(i\rho).$$

Hence, the total solution is written as

$$w = A_1 J_o\left(\pm ar\sqrt{i}\right) + A_2 Y_o\left(\pm ar\sqrt{i}\right) + A_3 J_o\left(\pm ar\sqrt{-i}\right) + A_4 Y_o\left(\pm ar\sqrt{-i}\right).$$

This equation can be written as (Hetenyi 1964)

$$w = C_1 Z_1(ar) + C_2 Z_2(ar) + C_3 Z_3(ar) + C_4 Z_4(ar) \tag{8.18}$$

where the functions Z_1 to Z_4 are modified Bessel functions given in Appendix B.

Example 8.4
A tubesheet in a heat exchanger shown in Figure 8.9 is subjected to edge load, Q_o, caused by the difference in expansion between the supporting tubes and the cylindrical shell. Find the expression for the deflection of the tubesheet due to force Q_o if the tubesheet is assumed simply supported at the edges.

Solution
From Eq. (8.18),

$$w = C_1 Z_1(ar) + C_2 Z_2(ar) + C_3 Z_3(ar) + C_4 Z_4(ar). \tag{1}$$

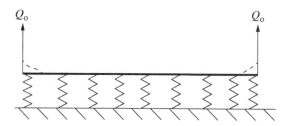

Figure 8.9   Plate on elastic foundation with edge load

The first constant is determined from the boundary condition at $r = 0$, where the slope dw/dr is equal to zero due to symmetry. Hence,

$$\frac{dw}{dr}\bigg|_{r=0} = C_1 a Z_1'(ar) + C_2 a Z_2'(ar) + C_3 a Z_3'(ar) + C_4 a Z_4'(ar).$$

From Figure B.3, the quantity $Z_4'(0)$ approaches infinity as r approaches zero. Hence, C_4 must be set to zero. The second constant is determined from the boundary at $r = 0$ where the shearing force, Q, is zero due to symmetry. The shearing force is expressed as

$$Q = D\left(\frac{d^3 w}{dr^3} + \frac{1}{r}\frac{d^2 w}{dr^2} - \frac{1}{r^2}\frac{dw}{dr}\right). \tag{2}$$

The derivatives of the first term in Eq. (1) are

$$\frac{dw}{dr} = C_1 a Z_1'(ar)$$

$$\frac{d^2 w}{dr^2} = C_1 a^2 Z_1''(ar)$$

or, from Appendix B,

$$\frac{d^2 w}{dr^2} = C_1\left[a^2 Z_2(ar) - \frac{a}{r}Z_1'(ar)\right].$$

The third derivative is

$$\frac{d^3 w}{dr^3} = C_1\left[a^3 Z_2'(ar) - \frac{a^2}{r}Z_1''(ar) + \frac{a}{r^2}Z_1'(ar)\right]$$

or, from Appendix B,

$$\frac{d^3 w}{dr^3} = C_1\left\{a^3 Z_2'(ar) - \frac{1}{r}\left[a^2 Z_2(ar) - \frac{a}{r}Z_1'(ar)\right] + \frac{a}{r^2}Z_1'(ar)\right\}.$$

Substituting these expressions into Eq. (2) yields

$$Q = D\left[C_1 a^3 Z_2'(ar)\right].$$

The derivatives of Z_2 and Z_3 in Eq. (1) are similar to those for Z_1. Thus, the total expression for Q in Eq. (2) becomes

$$Q = D\left[C_1 a^3 Z_2'(ar) - C_2 a^3 Z_1'(ar) + C_3 a^3 Z_4'(ar)\right].$$

At $r = 0$, $Q = 0$ due to symmetry. From Figure B.3, Z_1' and Z_2' have a finite value at $r = 0$ while Z_4' approaches ∞. Hence, C_3 must be set to zero. At the boundary condition $r = a$, we have

$$M_r|_{r=a} = 0 = -D\left(\frac{d^2w}{dr^2} + \frac{\mu}{r}\frac{dw}{dr}\right) \tag{3}$$

and

$$Q|_{r=a} = Q_o = D\left(\frac{d^3w}{dr^3} + \frac{1}{r}\frac{d^2w}{dr^2} - \frac{1}{r^2}\frac{dw}{dr}\right). \tag{4}$$

Substituting Eq. (1) into Eq. (3) yields

$$C_2 = C_1 \frac{\alpha Z_2(\alpha a) - \dfrac{1-\mu}{a} Z_1'(\alpha a)}{\alpha Z_1(\alpha a) + \dfrac{1-\mu}{a} Z_2'(\alpha a).} \tag{5}$$

And from Eqs. (4) and (5), we get

$$C_1 = \frac{-Q_o \alpha}{k_o F}\left[Z_1(\alpha a) + \frac{1-\mu}{\alpha a}Z_2'(\alpha a)\right]$$

where

$$F = Z_1(\alpha a)Z_2'(\alpha a) - Z_1'(\alpha a)Z_2(\alpha a) + \frac{1-\mu}{\alpha a}\left[Z_1'^2(\alpha a) + Z_2'^2(\alpha a)\right].$$

Substituting the expression for C_1 into Eq. (5) gives

$$C_2 = \frac{-Q_o \alpha}{k_o F}\left[Z_2(\alpha a) - \frac{1-\mu}{\alpha a}Z_1'(\alpha a)\right].$$

With C_1 and C_2 known, and $C_3 = C_4 = 0$, Eq. (1) can be solved for moments and shears throughout the plate.

8.4 Plates with Variable Boundary Conditions

In many structures such as large oil storage tanks (Figure 8.10a), the surface pressure above the contents causes an uplift force in the cylindrical shell. This force is normally transferred to the foundation through the anchor bolts. Many tanks, however, are not anchored to the foundation, especially in earthquake zones, to avoid damage to the tanks and their attachments. In such cases the uplift force due to surface pressure and earthquake loads is transferred to the base plate as shown in Figure 8.10b. The edge of the plate tends to lift up and the rest of the plate

(a)   (b)

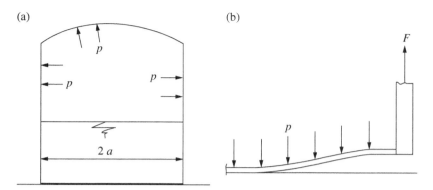

Figure 8.10   Uplift force in a flat bottom tank: (a) flat bottom tank with internal pressure and (b) uplift of the base plate

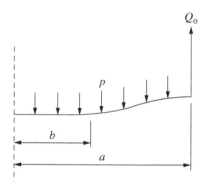

Figure 8.11   Uplift force Q

is kept in place by the pressure of the tank contents. The deflection of such a plate is obtained from (Jawad 2004) as

$$w = A + Br^2 + C\ln r + Fr^2 \ln r + \frac{pr^4}{64D} \tag{8.19}$$

where the constants A, B, C, and F are determined from the boundary conditions.

   At $r = a$, two boundary conditions can be specified. The first is the uplift force Q_o. The other boundary condition is obtained by specifying either M_r or θ. The other boundary conditions are obtained from Figure 8.11 by assuming an unknown dimension $r = b$ at which the following boundary conditions are satisfied:

$$w = 0, \quad \frac{dw}{dr} = 0, \quad M_r = 0.$$

These three boundary conditions plus the two at $r = a$ are used to solve the unknowns b, A, B, C, and F. As the constants A, B, C, and F are functions of b, the five equations obtained from the boundary conditions cannot be solved directly. A practical solution, however, can easily be obtained by using a spreadsheet program that increments various values of b until a solution is obtained.

Example 8.5

The tank shown in Figure 8.12 is subjected to an earthquake motion that results in an upward force at the cylinder-to-plate junction of 234 lbs/inch. Determine the maximum stress in the bottom plate and the maximum uplift. Assume the shell-to-plate junction to have zero rotation due to the cylindrical shell being substantially thicker than the base plate. Let $\mu = 0.3$ and $E = 29,000$ ksi.

Solution

At $r = b$ the deflection, w, is zero and Eq. (8.19) becomes

$$A + Bb^2 + C \ln b + Fb^2 \ln b = \frac{-pb^4}{64D}. \tag{1}$$

At $r = b$ the slope $dw/dr = 0$. Equation (8.19) gives

$$2Bb + C/b + Fb(2 \ln b + 1) = \frac{-4pb^3}{16D}. \tag{2}$$

At $r = b$, $M_r = 0$ and

$$M_r = -D \left(\frac{d^2 w}{dr^2} + \frac{\mu \, dw}{r \, dr} \right) = 0.$$

Substituting w into this equation gives

$$3.7143Bb^2 - C - Fb^2 (3.7143 \ln b + 4.7143) + \frac{4.1743pb4}{16D} = 0. \tag{3}$$

At $r = a$, $dw/dr = 0$ and

$$2Ba + C/a + Fa(2 \ln a + 1) = \frac{-4pa^3}{64D}. \tag{4}$$

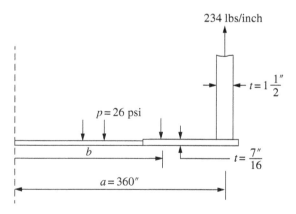

Figure 8.12   An API flat bottom tank

At $r = a$, $Q = Q_o$ and

$$Q_o = D\left(\frac{d^3w}{dr^3} + \frac{1}{r}\frac{d^2w}{dr^2} - \frac{1}{r^2}\frac{dw}{dr}\right),$$

which gives

$$F = \frac{Q_o a}{4D} - \frac{pa^2}{8D}. \tag{5}$$

Combining Eqs. (2), (3), (4), and (5) results in an equation that relates the unknown quantity b to the known loads Q_o and p. By placing all terms on one side of the equation, a computer program can be written to increment the quantity b until a solution is found that satisfies this equation. Using such a program results in a value of b of 346.2 inches. With this dimension known, the constants A, B, C, and F can now be determined and are given by

$$A = 1.1347 \times 10^{11} \qquad B = 2.5450 \times 10^6$$
$$C = -2.4615 \times 10^{10} \quad F = -4.0014 \times 10^5.$$

The maximum moment, M_r, is found to be 787.9 inch-lbs/inch and the maximum deflection at the edge as 0.02 inch:

$$\text{Maximum stress in bottom plate} = 6M/t^2 = 6 \times 787.9/0.4375^2$$

$$= 24,700 \, \text{psi}.$$

8.5 Design of Circular Plates

The references cited in Chapter 7 for rectangular plates also contain numerous tables for calculating maximum stress and deflection in circular plates of uniform thickness subjected to various loading and boundary conditions.

Circular plates are used as end closures in many shell structures such as reactors, heat exchangers, and distillation towers. Discussion of the interaction of circular plates with various shells will be discussed in later chapters. Table 8.1 lists a few loading conditions that will be utilized later when the interaction of plate and shell components is considered.

An approximate deflection and stress in perforated circular plates are obtained from the theoretical analysis of solid plates modified to take into consideration the effect of the perforations. One procedure that is commonly used is given in the ASME Code, Section VIII-2, Appendix 4. The code uses equivalent values of Poisson's ratio and modulus of elasticity in the theoretical equations for the deflection and stress of solid plates to obtain approximate values for perforated plates. The equivalent values are functions of the pitch and diameter of the perforations. The procedure is based, in part, on O'Donnell's work (O'Donnell and Langer, 1962).

The design of heat exchangers is based on Eq. (8.18) and its solution as shown in Example 8.4. Many codes and standards such as ASME-VIII and TEMA (2007) simplify the solution to a set of curves and equations suitable for design purposes.

Problems

8.1 The double wall storage tank is covered by a flat roof as shown. Find the moments in the roof due to an applied uniform load p, and draw the M_r and M_t diagrams. The attachment of the roof to the tank is assumed simply supported. Let $\mu = 0.3$.

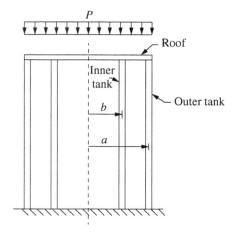

Problem 8.1 Double wall storage tank

8.2 Stainless steel baffles are attached to a vessel that has an agitator shaft. The attachment of the baffles to the vessel is assumed fixed, and the uniform pressure due to agitator rotation is 20 psi. What are the maximum values of M_r and M_t, and where do they occur? What is the maximum deflection at point b? Let $E = 27,000,000$ psi and $\mu = 0.29$. Also, if the baffles are assumed as fixed cantilevered beams, what will the maximum moment be, and how does it compare to M_r and M_t?

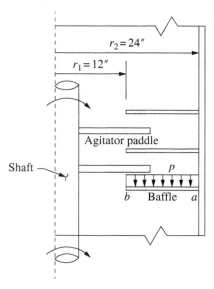

Problem 8.2 Baffles in a cylindrical vessel

8.3 A pan is made of aluminum and is full of water. If the edge of the bottom plate is assumed fixed, what is the maximum stress due to the exerted water pressure? Let $\gamma = 62.4$ pcf, $t = 0.030$ inch, $E = 10{,}200$ ksi, and $\mu = 0.33$. What is the maximum deflection?

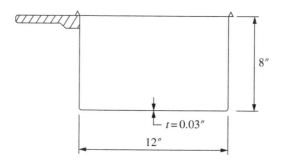

Problem 8.3   Aluminum pan

8.4 The pan in Problem 8.3 is empty and is at a temperature of 100°F. What is the thermal stress in the bottom plate if the bottom surface of the bottom plate is subjected to a temperature of 160°F and the top surface is subjected to a temperature of 40°F? Let the coefficient of expansion be 13.5×10^{-6} inches/inch/°F.

8.5 Find the expressions for the moments in the uniformly loaded circular plate.

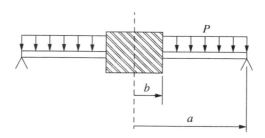

Problem 8.5   Plate with a rigid central attachment

8.6 Find the expressions for the moments in the circular plate with a central hole.

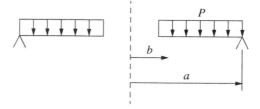

Problem 8.6   Plate with a central hole

8.7 Find the expressions for the moments in the circular plate with applied edge moments.

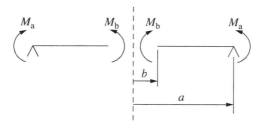

Problem 8.7   Plate with inner and outer moments

8.8 Show that the maximum deflection of a circular plate on an elastic foundation subjected to a concentrated load, F, in the center is given by the following expression when the radius of the plate is assumed infinitely large:

$$w_{max} = \frac{F}{8D\alpha^2}.$$

8.9 What are the values of C_1 and C_2 in Example 8.4 if the shear force Q_o is replaced by a bending moment M_o?

8.10 Find the stress in the bottom floor plate of the oil tank shown. Let $E = 200,000$ MPa and $\mu = 0.30$.

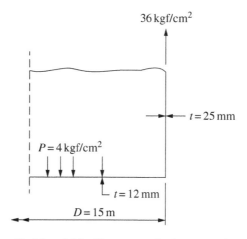

Problem 8.10   Forces on a flat bottom tank

8.11 Determine the maximum stress in the two cylindrical shells and in the 2-inch circular partition plate due to a pressure in the top compartment of 200 psi. Let $E = 25,000$ ksi and $\mu = 0.25$.

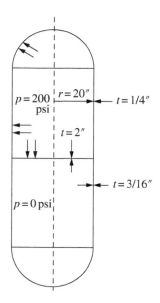

Problem 8.11   Cylinder shell with partition plate

8.12 Find the discontinuity forces at the cone to flat plate junction. Let $p = 60$ psi, $E = 20{,}000$ ksi, $\mu = 0.25$, and $\alpha = 36.87°$.

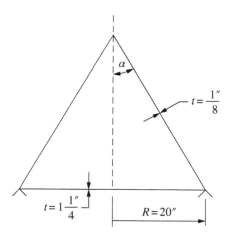

Problem 8.12   Conical shell attached to a circular plate

9

Approximate Analysis of Plates

9.1 Introduction

The analysis of rectangular plates having concentrated loads, free edges, or column supports is exceedingly complicated when using the classical theory discussed in Chapters 7 and 8. Similar difficulty is encountered in circular plates with concentrated loads as well as triangular plates with various loading and boundary conditions. The analysis commonly used to solve such cases is based on the plastic yield line theory, which is discussed in this chapter.

9.2 Yield Line Theory

The yield line theory is a powerful tool for solving many complicated plate problems where an exact elastic solution is impractical to obtain and an approximate solution is acceptable. It is best suited for plates with free boundary conditions and concentrated loads. The theory is based on the assumption that the stress–strain diagram of the material can be idealized as shown in Figure 9.1a. At point A on the elastic stress–strain diagram, the stress distribution across a plate of thickness t is as shown in Figure 9.1b. For a plate under external moment M, with no in-plane forces, the equilibrium equation for external and internal moments is

$$M = (\sigma_y)\left(\frac{t}{2}\right)\left(\frac{1}{2}\right)\left(\frac{2t}{3}\right)$$

or

$$\sigma_y = \frac{6M}{t^2}, \tag{9.1}$$

Stress in ASME Pressure Vessels, Boilers, and Nuclear Components, First Edition. Maan H. Jawad.
© 2018, The American Society of Mechanical Engineers (ASME), 2 Park Avenue,
New York, NY, 10016, USA (www.asme.org). Published 2018 by John Wiley & Sons, Inc.

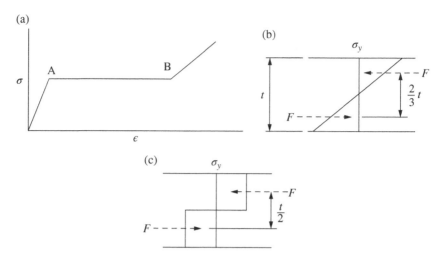

Figure 9.1 Elastic and plastic stress distribution: (a) Elastic-perfectly-plastic stress-strain diagram; (b) elastic stress distribution; and (c) plastic stress distribution

which is the basic relationship for bending stress of an elastic plate. For design purposes the yield stress is divided by a factor of safety to obtain an allowable stress.

As the load increases, the outer fibers of the plate are strained past the yield point A in Figure 9.1a. As the strain approaches point B, the stress distribution becomes, for all practical purposes, as shown in Figure 9.1c. Summation of internal and external moments gives

$$M_p = (\sigma_y)\left(\frac{t}{2}\right)\left(\frac{t}{2}\right)$$

or

$$\sigma_y = \frac{4M_p}{t^2}. \tag{9.2}$$

Equation (9.2) is the basic equation for the bending of a plate in accordance with the plastic theory. For design purposes, the applied loads are multiplied by a load factor to obtain the design loads.

A comparison of Eqs. (9.1) and (9.2) indicates that there is a reduction factor of 1.5 on stress level for a given moment and thickness. The comparison also shows there is a reduction in thickness by the amount of $(1.5)^{1/2}$ for a given moment and stress level.

Example 9.1

The maximum bending moment in a circular plate subjected to applied loads is 3000 inch-lbs/inch. Calculate the required thickness using elastic and plastic methods. Use a safety factor of 2.0, a load factor of 2.0, and a yield stress of 36,000 psi.

Solution
a. Elastic analysis

$$\sigma_a = \text{allowable stress} = \frac{36,000}{2.0} = 18,000 \text{ psi}$$

Hence,

$$t = \sqrt{\frac{6M}{\sigma_a}} = \sqrt{\frac{6 \times 3000}{18,000}} = 1.0 \text{ inch.}$$

b. Plastic analysis

$$M_p = 3000 \times 2.0 = 6000 \text{ inch-lbs/inch}$$

and

$$t = \sqrt{\frac{4M_p}{\sigma_y}} = \sqrt{\frac{4 \times 6000}{36,000}} = 0.82 \text{ inch.}$$

Hence, a savings of 22% in thickness is obtained by using plastic versus elastic analysis for the same factor of safety.

The yield line theory in plate analysis is very similar to the plastic hinge theory in beam analysis. Application of the plastic hinge theory in beams is illustrated by referring to the fixed beam shown in Figure 9.2a. The elastic moment diagram due to the concentrated load is shown in Figure 9.2b. The maximum elastic moment is $4FL/27$ and occurs at the right end support. As the load increases, the moment at the right end reaches M_p and a plastic hinge is developed at that location. However, because the moments at the left end and under the load are less than M_p, the beam can carry more load because it is still statistically determinate. Eventually the moment under the load reaches M_p and a plastic hinge is developed there also. However, the beam is still stable and more load can be applied to the beam until the moment at the left side reaches M_p, and the beam becomes unstable as shown in Figure 9.2c. At this instance the moment under the load, M_p, is equal to the moment, M_p, at the ends of the span. The magnitude of the moments can be determined by equating the external work to the internal work. The amount of external work is

$$\text{E.W.} = (F)(w),$$

while the internal work is given by

$$\text{I.W.} = M_p(\theta) + M_p(3\theta) + M_p(2\theta).$$

θ can be expressed as

$$\tan \theta \approx \theta = \frac{w}{(2/3)L}$$

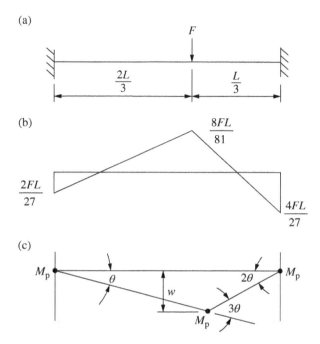

Figure 9.2 Beam with fixed ends: (a) fixed-ends beam, (b) moment distribution, and (c) plastic deflection

or

$$\theta = \frac{3w}{2L}.$$

Hence, the expression for internal work becomes

$$\text{I.W.} = 6M_p \frac{3w}{2L}.$$

Equating external and internal work gives

$$M_p = \frac{FL}{9}.$$

The ratio of M_p to M_e is 0.75. This coupled with the fact that $t_e = \sqrt{6M_e/S}$ while $t_p = \sqrt{4M_p/S}$ results in a net ratio of t_p to t_e of 0.7. Thus, a 30% savings in thickness is achieved by plastic analysis of this beam.

For plate analysis, the plastic hinges become yield lines. Also, axes of rotation develop in plates rather than points of rotation. Some of the properties of yield lines and axes of rotation are as follows:

1. In general, yield lines are straight.
2. Axes of rotation of a plate lie along lines of support.

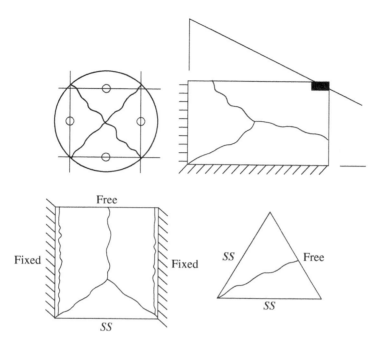

Figure 9.3   Yield lines for various plates

3. Axes of rotation pass over columns.
4. A yield line passes through the intersection of axes of rotation of adjacent plate segments.

Some illustrations of plates with various geometries, supports, and yield lines are shown in Figure 9.3. It must be noted that the failure mechanism method described here is an upper bound solution, and all failure mechanism patterns must be investigated in order to obtain a safe solution. However, the failure mechanisms for the class of problems discussed here have been verified experimentally and can thus be used for design purposes.

Example 9.2

Find the maximum plastic moment in a simply supported square plate subjected to a uniform load of intensity p.

Solution

The collapse mechanism of the square plate in Figure 9.4a consists of four yield lines. Section AA through the diagonal of the plate (Figure 9.4b) details the rotation of one of the yield lines. The plastic moment, M_p, at this yield line undergoes a rotation of 2α over the length od.

$$\text{External work} = \text{internal work}$$

$$(p) \, (\text{volume of pyramid}) = (M_p)(2\alpha)(\text{length})(4 \text{ yield lines})$$

$$(p)\left(L^2 \times w/3\right) = \left(M_p\right)(2\alpha)\left(\frac{\sqrt{2}L}{2}\right)(4).$$

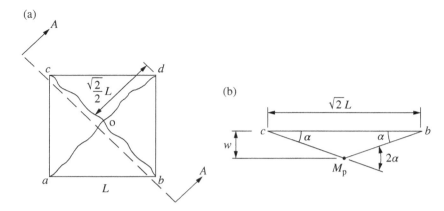

Figure 9.4   Yield lines in a simply supported plate: (a) plastic yield lines and (b) plastic deflection

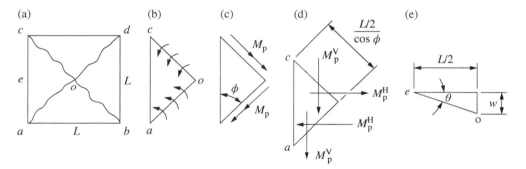

Figure 9.5   Vector representation of plastic moments: (a) plastic yield lines, (b) bending moments on a plate segment, (c) vector representation of the bending moments, (d) components of bending moments, and (e) deflected shape

For small deflection w, we let

$$\tan \alpha \approx \alpha = \frac{w}{\sqrt{2}L/2}.$$

We then get

$$M_{\mathrm{p}} = \frac{pL^2}{24}.$$

The analysis method discussed in Example 9.2 becomes cumbersome for complicated geometries. A more efficient method of formulating the expression for internal work is to use vector designations. We illustrate the method by referring to the square plate with the failure mechanism shown in Figure 9.5a. The applied moments in one panel (Figure 9.5b) can be designated by the vectors given in Figure 9.5c. The horizontal components of the vector moment (Figure 9.5d) cancel each other, while the vertical vectors are additive.

If we use the approach discussed in Example 9.2, then from sketches (d) and (e) of Figure 9.5, we get

$$\text{I.W. for one panel} = 2\left(M_p^v\right)\left(\frac{L/2}{\cos\phi}\right)(\theta)$$

$$= 2\left(M_p\cos\phi\right)\left(\frac{L/2}{\cos\phi}\right)\left(\frac{w}{L/2}\right)$$

$$= 2\left(M_p\right)w.$$

This same result can be obtained more efficiently by observing in sketches (d) and (e) of Figure 9.5 that the product of the quantity M_p times the slanted length is of the same magnitude as the product of the moment M_p applied along the edge L. Hence,

$$\text{I.W. for one panel} = \left(M_p\right)(L)(\theta)$$

$$= \left(M_p\right)(L)\left(\frac{w}{L/2}\right)$$

$$= 2\left(M_p\right)w.$$

As the rotation of the outside edge is obtained more easily than the rotation of the inner yield lines, the vector approach will be used in all subsequent discussions.

Example 9.3

Find the required thickness of a square plate fixed at the edges and subjected to a uniform load of 12 psi. Use a load factor of 2.0 and yield stress of 36 ksi.

Solution
From Figure 9.6a,

$$\text{E.W.} = \text{I.W.}$$

$$\frac{1}{3}\left(20^2\right)(2.0 \times 12)w = \text{work due to } M_p \text{ at internal yield lines}$$

$$+ \text{ work due to } M_p \text{ at edges}$$

and from Figure 9.6b,

$$3200\,w = 4\left(M_p\right)(L)\left(\frac{w}{L/2}\right) + 4\left(M_p\right)(L)\left(\frac{w}{L/2}\right)$$

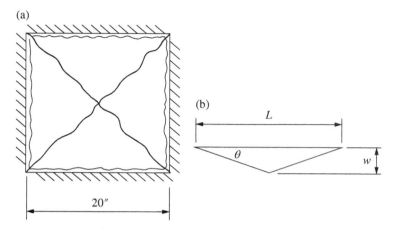

Figure 9.6 Square plate fixed along the edges: (a) plastic yield lines and (b) deflected shape

or

$$M_p = 200 \, \text{inch-lbs/inch}.$$

$$t = \sqrt{\frac{4M_p}{\sigma_y}} = \sqrt{\frac{4 \times 200}{36,000}} = 0.15 \, \text{inch}.$$

Example 9.4

Find M_p of the rectangular plate in Figure 9.7a due to uniform load p. The plate is simply supported.

Solution

The yield lines take the shape shown in Figure 9.7b. Distance x is unknown and must be determined:

$$\text{I.W.} = 2\left(M_p\right)(a)\left(\frac{w}{x}\right) + 2\left(M_p\right)(b)\left(\frac{w}{a/2}\right).$$

From Figure 9.7c,

$$\text{E.W.} = p(\text{E.W. I} + \text{E.W. II} + \text{E.W. III})$$

$$= p\left[2\left(\frac{ax}{2}\right)\left(\frac{1}{3}\right) + 2(b-2x)\left(\frac{a}{2}\right)\left(\frac{1}{2}\right) + 4(x)\left(\frac{a}{2}\right)\left(\frac{1}{2}\right)\left(\frac{1}{3}\right)\right]w.$$

Equating internal and external work and solving for M_p gives

$$M_p = p\left[\frac{\dfrac{ab}{2} - \dfrac{ax}{3}}{2\left(\dfrac{2b}{a} + \dfrac{a}{x}\right)}\right]. \tag{1}$$

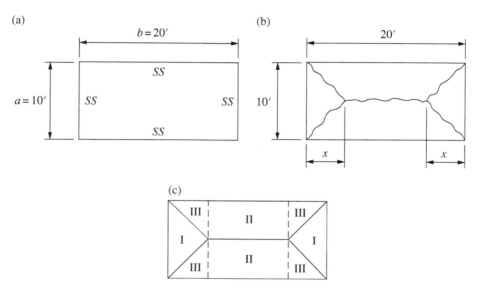

Figure 9.7 Simply supported rectangular plate: (a) rectangular plate, (b) plastic yield lines, and (c) designate of plate segments

Minimizing M_p by taking its derivative with respect to x and equating it to zero gives

$$x^2 + \frac{a^2}{b}x - \frac{3a^2}{4} = 0. \tag{2}$$

For $a = 10$ ft, and $b = 20$ ft, Eq. (2) gives $x = 6.51$ ft. From Eq. (1),

$$M_p = p\left(\frac{14,400 - 3124.8}{2(4.00 + 1.54)}\right)$$

$$= 1018p \text{ inch-lbs/inch.}$$

9.3 Further Application of the Yield Line Theory

The deflections obtained from the yield line theory are larger than those obtained from the elastic theory due to reduced thicknesses. This should be considered in applications where small deflections are critical to the performance of equipment such as flanges and other sealing components. Also, the yield line theory tends to give an upper bound solution. Accordingly, all possible yield line paths must be investigated in order to obtain a true solution. This is especially important in plates with free edges as illustrated in Example 9.5.

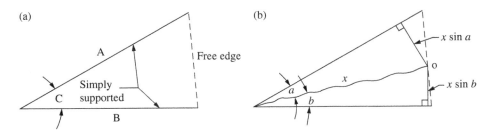

Figure 9.8 Plate with a free edge: (a) triangular plate and (b) plastic yield lines

Example 9.5
Find the maximum moment in a triangular section (Figure 9.8a) subjected to a uniform load p.

Solution
Assume the yield line to be as shown in Figure 9.8b at an angle a from side A. Also, assume w to be the deflection at point O. Then

$$C = a + b$$

$$\text{I.W.} = M_p(x\cos a)\frac{w}{x\sin a} + M_p(x\cos b)\frac{w}{x\sin b}$$

$$= M_p(\cot a + \cot b)w$$

$$\text{E.W.} = p\left[\frac{(A)(x\sin a)}{2}\frac{1}{3} + \frac{(B)(x\sin b)}{2}\frac{1}{3}\right]w$$

$$= \frac{px}{6}(A\sin a + B\sin b)w.$$

Equating internal and external work results in

$$M_p = \frac{px}{a}\frac{A\sin a + B\sin b}{\cot a + \cot b}.$$

Minimizing M_p with respect to a results in

$$a = \frac{C}{2},$$

which indicates that in any triangular plate with one edge free, the yield line always bisects the angle between the two simply supported edges. The plastic moment is given by

$$M_p = \frac{px}{6}\frac{A\sin(C/2) + B\sin(C/2)}{\cot(C/2) + \cot(C/2)},$$

where

$$x = \frac{K_1}{\sin(C/2)}\sin K_2$$

$$K_1 = 0.5\left(A^2 + B^2 - 2AB\cos C\right)^{1/2}$$

$$K_2 = 180 - \left(\frac{C}{2} + K_3\right)$$

$$K_3 = \sin^{-1}\left(\frac{2B}{2K_1}\sin\frac{C}{2}\right).$$

Example 9.6

Find the maximum moment in a uniformly loaded square plate (Figure 9.9a) with three sides simply supported and one side free.

Solution

a. Let the failure pattern be as shown in Figure 9.9b. Let maximum deflection w be at point 0. Then,

$$I.W. = 2M_p(L)\frac{w}{L/2} + M_p(L)\frac{w}{L-y}$$

$$= \left(4M_p + \frac{M_pL}{L-y}\right)w$$

$$E.W. = p\left[\frac{L(L-y)}{2}\frac{1}{3} + \frac{Ly}{2} + 2(L-y)\frac{L}{2}\frac{1}{2}\frac{1}{3}\right]w$$

$$= p\left[\frac{L(L-y)}{3} + \frac{Ly}{2}\right]w.$$

Equating internal and external work gives

$$M_p = \frac{pL}{6}\frac{2L^2 - Ly - y^2}{5L - 4y}.$$

Minimizing M_p with respect to y gives

$$y = 2.15L, \text{ which is discarded,}$$

or

$$y = 0.35L$$

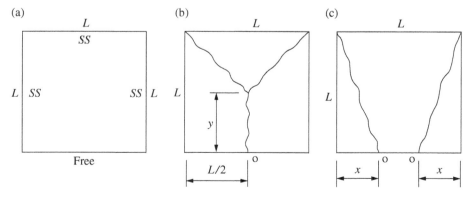

Figure 9.9 Square plate with one side free: (a) square plate, (b) first assumed pattern of plastic yield lines, and (c) second assumed pattern of plastic yield lines

and

$$M_p = \frac{pL^2}{14.14}.$$

b. Let the failure pattern be as shown in Figure 9.9c. Let maximum deflection w be along points 0–0. Then,

$$I.W. = 2M_pL\frac{w}{x} + 2M_px\frac{w}{x}$$

$$= \left(\frac{2M_pL}{x} + \frac{2M_px}{L}\right)w$$

$$E.W. = p\left[\frac{2Lx}{2}\frac{1}{3} + \frac{2Lx}{2}\frac{1}{3} + (L-2x)L\frac{1}{2}\right]w$$

$$= p\left[\frac{2Lx}{3} + \frac{(L-2x)L}{2}\right]w.$$

Equating internal and external work results in

$$M_p = \frac{pL}{12}\frac{3L^2x - 2x^2L}{L+x}.$$

Minimizing M_p with respect to x gives

$$x = -1.87L \text{ or } x = 0.54L, \text{ which are impossible.}$$

Thus, the first alternate (a) controls.

The yield line mechanism in circular plates subjected to uniform or concentrated loads has a radial pattern. The failure mode is conical in shape as illustrated in Example 9.7. This circular pattern must also be investigated for rectangular plates under concentrated loads.

Example 9.7

Find the moment in a simply supported circular plate due to (a) uniform load p and (b) concentrated load F in the middle.

Solution

a. From Figure 9.10 using w as the deflection in the middle,

$$I.W. = \int_0^{2\pi} M_pa\,d\phi\frac{w}{a}$$

$$= 2\pi M_pw$$

$$E.W. = (\pi a^2p/3)w$$

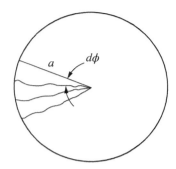

Figure 9.10 Yield lines in a circular plate

and

$$M_p = \frac{pa^2}{6}.$$

b. Again from Figure 9.10 using w as the deflection in the middle,

$$\text{E.W.} = F(w)$$

and

$$M_p = \frac{F}{2\pi}.$$

It is of interest to note that M_p in this case is independent of the radius.

Example 9.8
Find the moment in a square plate subjected to a concentrated load F in the middle if the plate is (a) simply supported on all sides or (b) fixed on all sides.

Solution
a. From Figure 9.11a, using a straight line mechanism with a deflection, w, in the middle,

$$4(M_p)(a)\frac{w}{a/2} = F(w)$$

$$M_p = 0.125F.$$

From Figure 9.11b, using a circular mechanism with a deflection, w, in the middle,

$$2(2\pi M_p)w = F(w)$$

$$M_p = 0.08F.$$

Hence, the straight line mechanism controls.

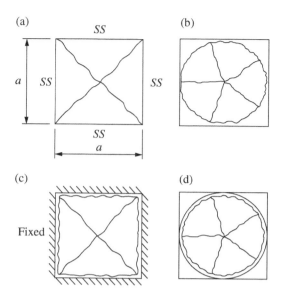

Figure 9.11 Yield lines in a square plate: (a) first assumed pattern of plastic yield lines in a simply supported plate; (b) second assumed pattern of plastic yield lines in a simply supported plate; (c) first assumed pattern of plastic yield lines in a fixed plate, and (d) second assumed pattern of plastic yield lines in a fixed plate

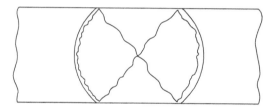

Figure 9.12 Fanlike yield lines in long narrow plates

b. From Figure 9.11c, using a straight line mechanism with a deflection, w, in the middle,

$$2\left[4(M_\mathrm{p})(a)\frac{1}{a/2}\right]w = F(w)$$

$$M_\mathrm{p} = 0.063F.$$

From Figure 9.11d, using a circular mechanism with a deflection, w, in the middle,

$$2(2\pi M_\mathrm{p})w = F(w)$$

$$M_\mathrm{p} = 0.08F$$

and the circular mechanism controls.

Another failure mechanism that occurs in long narrow plates is shown in Figure 9.12 and should be investigated together with other failure modes.

9.4 Design Concepts

The plastic theory is used as an approximation for determining maximum moments in plates that cannot be solved easily by the classical plate theory due to complex geometry, boundary conditions, or applied loading. In many structures this approximation is adequate for design purposes. When a more accurate analysis is needed, the plastic theory is used first as an approximation followed by a more rigorous analysis such as the finite element method. Table 9.1 lists the maximum plastic moment for some frequently encountered plates with various loading and boundary conditions. It should be noted that the deflection due to plastic design is larger than that obtained from the elastic theory due to reduced thickness. Thus, extra precaution must be given to the design of components that cannot tolerate large deflections such as cover plates in flanged openings.

Equation (9.2) for plastic bending of a plate is used by numerous international codes to establish an upper limit on the allowable bending stress values. The ratio obtained by dividing Eq. (9.1) by Eq. (9.2) is 1.5 and is referred to as the shape factor. It indicates that for a given bending moment, plastic analysis of plates results in a stress level that is 50% lower than that determined from the elastic theory for the same factor of safety. Accordingly, many standards such as the ASME Boiler and Pressure Vessel Code use an allowable stress for plates in bending that is 50% higher than the tabulated allowable membrane stress value.

The finite difference method is an excellent design tool for solving plates that are irregular in shape, have internal supports, or have complex boundary conditions. The solution is easily obtained by solving a set of simultaneous equations, and the accuracy of the answer can readily be verified by decreasing the mesh size and comparing the results.

Table 9.1 Plastic bending moments in various plates

Case	Maximum moment
1. Uniform load, p, simply supported edge	$M_p = \dfrac{pr^2}{6}$
2. Uniform load, p, fixed edge	$M_p = \dfrac{pr^2}{12}$
3. Concentrated load, F, in the middle simply supported edge	$M_p = \dfrac{F}{2\pi}$
4. Concentrated load, F, in the middle fixed edge	$M_p = \dfrac{F}{4\pi}$

(continued overleaf)

Table 9.1   (*continued*)

Case	Maximum moment
5. Uniform load, p, simply supported edges	$M_p = \dfrac{pL^2}{24}$
6. Uniform load, p, fixed edges	$M_p = \dfrac{pL^2}{48}$
7. Concentrated load, F, in the middle simply supported edges	$M_p = \dfrac{F}{8}$
8. Concentrated load, F, in the middle fixed edges	$M_p = \dfrac{F}{4\pi}$
9. Uniform load, p, simply supported edges	$M_p = \dfrac{pa^2}{72}$
10. Uniform load, p, fixed edges	$M_p = \dfrac{pa^2}{144}$
11. Concentrated load, F, at the centroid simply supported edges	$M_p = \dfrac{F}{6\sqrt{3}}$
12. Concentrated load, F, at the centroid fixed edges	$M_p = \dfrac{F}{12\sqrt{3}}$

Problems

9.1 Find M_p due to a uniform load on the simply supported hexagon plate.

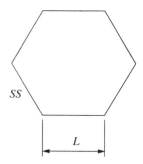

Problem 9.1 Hexagonal plate

9.2 Find M_p in the skewed plate due to uniform load p.

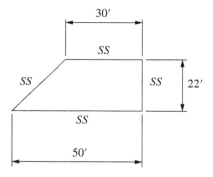

Problem 9.2 Trapezoidal plate

9.3 Determine M_p in the triangular plate due to uniform load p. The plate is fixed at all edges.

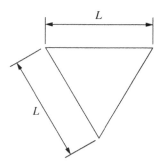

Problem 9.3 Triangular plate

9.4 Find M_p in the rectangular base plate due to uniform load p.

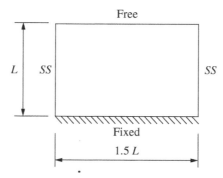

Problem 9.4   Rectangular plate with various boundary conditions

9.5 Find M_p in Problem 9.4 if a concentrated load F is applied at the middle of the free edge.
9.6 The internal vessel tray is assumed fixed at the outer edge. Determine M_p due to a concentrated load in the middle.

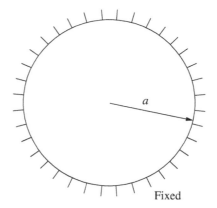

Problem 9.6   Circular plate fixed at edge

9.7 Determine M_p in the triangular plate shown due to uniform load p.

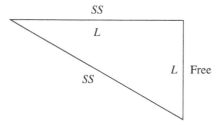

Problem 9.7   Triangular plate

9.8 Find M_p in the rectangular plate with two adjacent sides fixed and the opposite corner supported by a column. The plate is uniformly loaded with 75 psf.

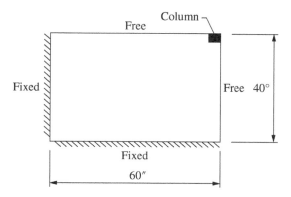

Problem 9.8 Plate supported by a column at one corner

10

Buckling of Plates

10.1 Circular Plates

When a thin elastic plate is subjected to compressive in-plane axial loads, in conjunction with small applied lateral loads or imperfections in the plate, the in-plane deflections increase gradually with an increase in the applied loads up to a certain critical point. Beyond this point a slight increase in axial loads causes a large and sudden increase in the deflection. This phenomenon, called buckling, is the subject of this chapter for circular and rectangular plates. A more comprehensive treatment of this subject is given by Timoshenko and Gere (1961), Bloom and Coffin (2001), Szilard (1974), and Iyengar (1988).

The differential equation for the bending of a circular plate subjected to lateral loads, p, is obtained from Eq. (8.11) as

$$r^2 \frac{d^2\phi}{dr^2} + r\frac{d\phi}{dr} - \phi = -\frac{Qr^2}{D}.$$

where
$D = Et^3/12(1-\mu^2)$
$E =$ modulus of elasticity
$Q =$ shear
$r =$ radius of plate
$t =$ thickness
$\phi =$ angle as shown in Figure 10.1.
$\mu =$ Poisson's ratio

When in-plane forces N_r are applied as shown in Figure 10.1, and the lateral loads, p, are reduced to zero, then the corresponding value of Q is

$$Q = N_r\phi.$$

Stress in ASME Pressure Vessels, Boilers, and Nuclear Components, First Edition. Maan H. Jawad.
© 2018, The American Society of Mechanical Engineers (ASME), 2 Park Avenue,
New York, NY, 10016, USA (www.asme.org). Published 2018 by John Wiley & Sons, Inc.

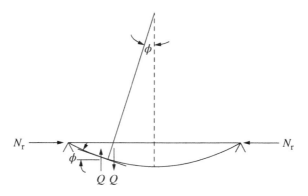

Figure 10.1 In-plane load in a circular plate

Letting

$$A^2 = \frac{N_r}{D} \tag{10.1}$$

the differential equation becomes

$$r^2 \frac{d^2\phi}{dr^2} + r\frac{d\phi}{dr} - \left(r^2 A^2 - 1\right)\phi = 0.$$

Defining

$$x = Ar \quad \text{and} \quad dx = A\,dr \tag{10.2}$$

we get

$$x^2 \frac{d^2\phi}{dx^2} + x\frac{d\phi}{dx} + \left(x^2 - 1\right)\phi = 0. \tag{10.3}$$

The solution of this equation is in the form of a Bessel function. From Eq. (B.3) of Appendix B,

$$\phi = C_1 J_1(x) + C_2 Y_1(x) \tag{10.4}$$

at $r = 0$, $Y_1(x)$ approaches infinity. Hence, C_2 must be set to zero and Eq. (10.4) becomes

$$\phi = C_1 J_1(x). \tag{10.5}$$

For a fixed boundary condition, $\phi = 0$ at $r = a$ and a nontrivial solution of Eq. (10.5) is

$$J_1(x) = 0$$

or from Table B.1, $x = 3.83$, and Eqs. (10.1) and (10.2) give

$$\sqrt{N_r/D}(a) = 3.83$$

or

$$N_{cr} = \frac{14.67D}{a^2}. \tag{10.6}$$

For a simply supported plate, the moment at the boundary $r = a$ is zero and Eqs. (10.5) and (8.4) give

$$N_{cr} = \frac{4.20D}{a^2}. \tag{10.7}$$

Equations (10.6) and (10.7) are for the critical buckling load of circular plates with fixed and simply supported boundary conditions, respectively. It is of interest to note that contrary to column buckling where the column is rendered ineffective in carrying any further loads subsequent to buckling, a simply supported circular plate is capable of carrying additional loads in the postbuckling phased due to in-plane biaxial membrane stress (Bloom and Coffin 2001). Postbuckling analysis of such plates (Sherbourne 1961) results in an equation of the form

$$\frac{N_u}{N_{cr}} = 1 + 0.241 \left(\frac{w}{t}\right)^2 \tag{10.8}$$

where
N_u = ultimate axial load
N_{cr} = critical buckling load
t = plate thickness
w = plate deflection

This equation is independent of the ratio r/t and is limited to $w/t < 3.0$. Values greater than 3.0 result in plastic strains in the plate.

Example 10.1
What is the required thickness of a simply supported circular plate subjected to a lateral pressure of 2 psi and in-plane compressive force of 100 lbs/inch if $a = 29$ inches, $\mu = 0.31$, $E = 30,000$ ksi, allowable stress in bending = 10,000 psi, and factor of safety (F.S.) for buckling = 3.0?

Solution
From Example 8.1,

$$M = \frac{Pa^2}{16}(3 + \mu) = 348.0 \text{ inch-lbs/inch}$$

$$t = \sqrt{6 \times 348/10,000} = 0.46 \text{ inch}.$$

Try $t = 0.50$ inch.

$$\text{Actual bending stress} = \frac{6 \times 348}{0.5^2} = 8350\,\text{psi}$$

$$\text{Actual compressive stress} = \frac{100}{0.5} = 200\,\text{psi}$$

$$D = \frac{30,000,000 \times 0.5^3}{12(1-0.31^2)} = 345,720\,\text{inch-lbs.}$$

From Eq. (10.7),

$$\sigma_{cr} = \frac{4.20 \times 345,720}{29^2 \times 0.5} = 3450\,\text{psi.}$$

Allowable compressive stress $= 3450/3 = 1150$ psi.
    Using the interaction equation

$$\frac{\text{Actual bending stress}}{\text{Allowable bending stress}} + \frac{\text{Actual compressive stress}}{\text{Allowable compressive stress}} \leq 1.0$$

gives

$$\frac{8350}{10,000} + \frac{200}{1150} = 0.83 + 0.17 = 1.0 \,(\text{acceptable stress})$$

Use $t = 0.50$ inch.
    The governing differential equation for the buckling of circular plates with a central hole, that is, annular plate, is derived in much the same way as that for solid circular plates. However, the general differential equation (Bloom and Coffin 2001) for an annular plate is more complicated than a solid plate. The equation is similar to the nonsymmetric bending (Jawad 2004) of a plate with the exception of having applied axial in-plane loads N_r and tangential in-plane loads N_θ rather than lateral loads p. Shearing forces $N_{\theta r}$ are assumed to be zero. The equation is expressed as

$$D\nabla^4 w = -N_r \frac{\partial^2 w}{\partial r^2} - N_\theta \left(\frac{1}{r}\frac{\partial^2 w}{\partial r \partial \theta} + \frac{1}{r^2}\frac{\partial^2 w}{\partial \theta^2} \right) \tag{10.9}$$

where $\nabla^4 w$ is given by

$$\nabla^4 w = \frac{\partial^4 w}{\partial r^4} + \frac{2}{r}\frac{\partial^3 w}{\partial r^3} - \frac{1}{r^2}\frac{\partial^2 w}{\partial r^2} + \frac{1}{r^3}\frac{\partial w}{\partial r} + \frac{2}{r^2}\frac{\partial^4 w}{\partial r^2 \partial \theta^2} - \frac{2}{r^3}\frac{\partial^3 w}{\partial \theta^2 \partial r} + \frac{4}{r^4}\frac{\partial^2 w}{\partial \theta^2} + \frac{1}{r^4}\frac{\partial^4 w}{\partial \theta^4}. \tag{10.10}$$

The solution of Eq. (10.9) for specific plate geometry with the appropriate boundary conditions results in four simultaneous equations involving Bessel functions. Setting the determinate of the coefficients of these four equations to zero and solving the resultant polynomial provides

Table 10.1 Critical buckling coefficient, k, of an annular plate with a free inside edge

b/a	k Values	
	Simply supported outer edge	Fixed outer edge
0.0	4.20	14.64
0.2	3.60	13.32
0.4	2.76	17.76
0.6	2.28	Note 1
0.8	2.04	Note 1

$N_{cr} = kD/a^2$

where

a = outer radius
b = inside radius
$D = Et^3/12(1 - \mu^2)$
E = modulus of elasticity
k = constant
N_{cr} = critical load, force/length
t = thickness
μ = Poisson's ratio

Note 1: The critical load increases exponentially (Timoshenko and Gere 1961).

the eigenvalues associated with buckling of the specific plate and boundary conditions. The critical buckling load is obtained from the lowest eigenvalue. Table 10.1 lists the critical buckling load for an annular plate subjected to in-plane compressive loads at the outside edge. The inner edge is free and the outer edge is either simply supported or fixed.

10.2 Rectangular Plates

Buckling of rectangular plates is most commonly caused by in-plane shear or in-plane axial loads in the x- and y-directions. In this and the following sections, classical as well as numerical methods are presented for solving these loading conditions (Gerard and Becker 1957a, 1957b). Other loading conditions due to temperature gradients and support settlements are beyond the scope of this book.

The differential equation for the bending of a rectangular plate with lateral load, p, is given by Eq. (7.26) as

$$\nabla^4 w = \frac{p}{D}. \tag{10.11}$$

If the plate is additionally loaded in its plane (Figure 10.2a), then summation of forces in the x-direction gives

$$N_x dy + N_{yx} dx - \left(N_x + \frac{\partial N_x}{\partial x} dx \right) dy - \left(N_{yx} + \frac{\partial N_{yx}}{\partial y} dy \right) dy = 0$$

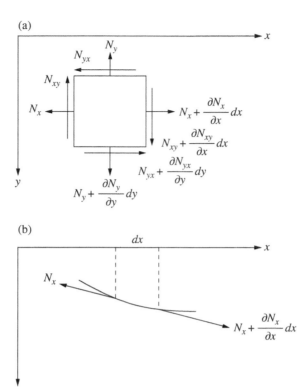

Figure 10.2   In-plane loads in a rectangular plate: (a) infinitesimal element and (b) membrane force N_x

or

$$\frac{\partial N_x}{\partial x} + \frac{\partial N_{yx}}{\partial y} = 0. \tag{10.12}$$

Similarly, summation of forces in the y-direction gives

$$\frac{\partial N_y}{\partial y} + \frac{\partial N_{xy}}{\partial x} = 0. \tag{10.13}$$

Summation of forces in the z-direction (Figure 10.2b) gives the following for N_x:

$$-N_x dy \frac{\partial w}{\partial x} + \left(N_x + \frac{\partial N_x}{\partial x} dx \right) \left(\frac{\partial w}{\partial x} + \frac{\partial}{\partial x}\left(\frac{\partial w}{\partial x} \right) dx \right) dy$$

which reduces to

$$N_x \frac{\partial^2 w}{\partial x^2} dx dy + \frac{\partial N_x}{\partial x} \frac{\partial w}{\partial x} dx dy. \tag{10.14}$$

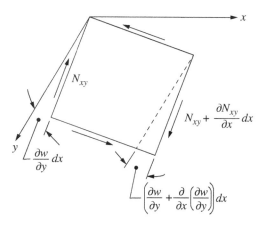

Figure 10.3   In-plane shear loads

Similarly, for N_y,

$$N_y \frac{\partial^2 w}{\partial y^2} dx \, dy + \frac{\partial N_y}{\partial y} \frac{\partial w}{\partial y} dx \, dy. \tag{10.15}$$

For N_{xy} from Figure 10.3,

$$-\left(N_{xy} dy \frac{\partial w}{\partial y}\right) + \left(N_{xy} + \frac{\partial N_{xy}}{\partial x} dx\right)\left(\frac{\partial w}{\partial y} + \frac{\partial^2 w}{\partial x \partial y} dx\right) dy$$

or

$$N_{xy} \frac{\partial^2 w}{\partial x \partial y} dx \, dy + \frac{\partial N_{xy}}{\partial x} \frac{\partial w}{\partial y} dx \, dy. \tag{10.16}$$

Similarly for N_{yx}

$$N_{yx} \frac{\partial^2 w}{\partial x \partial y} dx \, dy + \frac{\partial N_{yx}}{\partial y} \frac{\partial w}{\partial x} dx \, dy. \tag{10.17}$$

The total sum of Eq. (10.11), which was obtained by summing forces in the z-direction, with Eqs. (10.14), (10.15), (10.16), and (10.17) gives the basic differential equation of a rectangular plate subjected to lateral and in-plane loads.

$$\nabla^4 w = \frac{1}{D}\left(p + N_x \frac{\partial^2 w}{\partial x^2} + N_y \frac{\partial^2 w}{\partial y^2} + 2N_{xy} \frac{\partial^2 w}{\partial x \partial y}\right). \tag{10.18}$$

It should be noted that Eqs. (10.12) and (10.13), which were obtained by summing forces in the x- and y-directions, were not utilized in Eq. (10.18). They are used to formulate large-deflection theory of plates that is beyond the scope of this book.

Another equation that is frequently utilized in buckling problems is the energy equation. The strain energy expression due to lateral loads, p, is given by

$$U = \frac{D}{2}\int_{A}\left\{\left(\frac{\partial^2 w}{\partial x^2} + \frac{\partial^2 w}{\partial y^2}\right)^2 - 2(1-\mu)\left[\frac{\partial^2 w}{\partial x^2}\frac{\partial^2 w}{\partial y^2} - \left(\frac{\partial^2 w}{\partial x \partial y}\right)^2\right]\right\}dx\,dy. \tag{10.19}$$

The strain energy expression for the in-plane loads is derived from Figure 10.4, which shows the deflection of a unit segment dx. Hence,

$$dx' = \sqrt{dx^2 - \left(\frac{\partial w}{\partial x}dx\right)^2}$$

$$\varepsilon_x = dx - dx' = \frac{1}{2}\left(\frac{\partial w}{\partial x}\right)^2 dx$$

or per unit length,

$$\varepsilon_x = \frac{1}{2}\left(\frac{\partial w}{\partial x}\right)^2.$$

Similarly,

$$\varepsilon_y = \frac{1}{2}\left(\frac{\partial w}{\partial y}\right)^2.$$

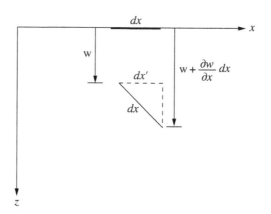

Figure 10.4   Deflection of a unit segment

It can also be shown that

$$\gamma_{xy} = \frac{\partial w}{\partial x}\frac{\partial w}{\partial y}.$$

Hence, the strain energy expression for the in-plane forces is given by

$$U = \int_A \left(N_x \varepsilon_x + N_y \varepsilon_y + N_{xy}\gamma_{xy}\right)dx\,dy$$

$$U = \frac{1}{2}\int_A \left[N_x\left(\frac{\partial w}{\partial x}\right)^2 + N_y\left(\frac{\partial w}{\partial y}\right)^2 + 2N_{xy}\frac{\partial w}{\partial x}\frac{\partial w}{\partial y}\right]dx\,dy.$$

(10.20)

The total strain energy expression for rectangular plates loaded laterally and in-plane is the summation of expressions (10.19) and (10.20). Thus,

$$U = \frac{D}{2}\int_A \left\{\left(\frac{\partial^2 w}{\partial x^2} + \frac{\partial^2 w}{\partial y^2}\right)^2 - 2(1-\mu)\left[\frac{\partial^2 w}{\partial x^2}\frac{\partial^2 w}{\partial y^2} - \left(\frac{\partial^2 w}{\partial x\partial y}\right)^2\right]\right\}dxdy$$

$$+ \frac{1}{2}\int_A \left[N_x\left(\frac{\partial w}{\partial x}\right)^2 + N_y\left(\frac{\partial w}{\partial y}\right)^2 + 2N_{xy}\frac{\partial w}{\partial x}\frac{\partial w}{\partial y}\right]dxdy.$$

(10.21)

The total potential energy of a system is given by

$$\prod = U - W$$

(10.22)

where W is external work. In order for the system to be in equilibrium, Eq. (10.22) must be minimized.

Example 10.2
Find the buckling stress of a simply supported rectangular plate, Figure 10.5, subjected to forces N_x.

Solution
Let the deflection be expressed as

$$w = \sum_{m=1}^{\infty}\sum_{n=1}^{\infty}A_{mn}\sin\frac{m\pi x}{a}\sin\frac{n\pi y}{b},$$

which satisfies the boundary conditions.

Substituting this expression into Eq. (10.21), and noting that the expression

$$2(1-\mu)\left[\frac{\partial^2 w}{\partial x^2}\frac{\partial^2 w}{\partial y^2} - \left(\frac{\partial^2 w}{\partial x\partial y}\right)^2\right] = 0,$$

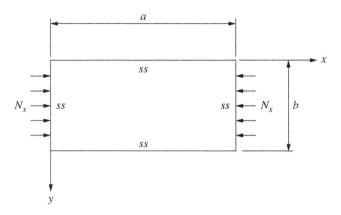

Figure 10.5   Rectangular plate subjected to force N_x

gives

$$U = \frac{D}{2} \int_0^b \int_0^a \sum_{m=1}^{\infty} \sum_{n=1}^{\infty} \left[A_{mn}^2 \left(\frac{m^2 \pi^2}{a^2} + \frac{n^2 \pi^2}{b^2} \right)^2 \sin^2 \frac{m\pi x}{a} \sin^2 \frac{n\pi y}{b} \right] dx\, dy$$

$$+ \frac{1}{2} \int_0^b \int_0^a \left[(-N_x) \sum_{m=1}^{\infty} \sum_{m=1}^{\infty} A_{mn}^2 \frac{m^2 \pi}{a^2} \times \sin^2 \frac{m\pi x}{a} \sin^2 \frac{n\pi y}{b} \right] dx\, dy$$

or

$$U = \frac{\pi^4 ab}{8} D \sum_{m=1}^{\infty} \sum_{n=1}^{\infty} A_{mn} \left(\frac{m^2}{a^2} + \frac{n^2}{b^2} \right)^2 - \frac{\pi^2 b}{8a} N_x \sum_{m=1}^{\infty} \sum_{n=1}^{\infty} m^2 A_{mn}^2. \tag{1}$$

Since there are no lateral loads, we can take the external work in the z-direction as zero and Eq. (10.22) becomes

$$\prod = U.$$

Solving Eq. (1) for

$$\frac{\partial \prod}{\partial A_{mn}} = 0$$

we get

$$N_{cr} = \frac{\pi^2 a^2 D}{m^2} \left(\frac{m^2}{a^2} + \frac{n^2}{b^2} \right)^2. \tag{2}$$

The smallest value of Eq. (2) is for $n = 1$. Also, if we substitute

$$\sigma_{cr} = \frac{N_{cr}}{t}$$

and

$$D = \frac{Et^3}{12(1-\mu^2)}$$

we get

$$\sigma_{cr} = \frac{\pi^2 E}{12(1-\mu^2)(b/t)^2} K \tag{3}$$

where

$$K = \left(\frac{m}{a/b} + \frac{a/b}{m} \right)^2. \tag{4}$$

A plot of Eq. (4) is shown in Figure 10.6 and shows that the minimum value of K is 4.0.
Equation (3) in Example 10.2 gives the critical elastic buckling stress in a simply supported rectangular plate. The minimum value when $k = 4.0$ is

$$\sigma_{cr} = \frac{\pi^2 E}{3(1-\mu^2)(b/t)^2}. \tag{10.23}$$

This equation can also be written as

$$\sigma_{cr} = K \left(\frac{D}{t} \right) \left(\frac{\pi}{b} \right)^2 \tag{10.24}$$

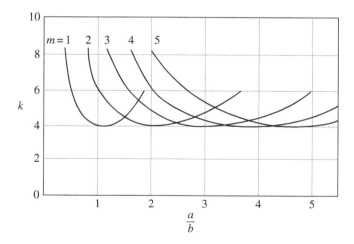

Figure 10.6   Plot of Eq. (4) in Example 10.2

where
b = short dimension of plate
$D = Et^3/12(1 - \mu^2)$
E = modulus of elasticity
K = 4.0
t = thickness of plate
σ_{cr} = critical buckling stress of plate

Theoretical postbuckling equations (Bloom and Coffin 2001) as well as experimental work has shown that the critical buckling stress in rectangular plates increases significantly from that of Eq. (10.24) as the buckling deflection increases. This is demonstrated in Figure 10.7 as a function of the ratio of plate deflection, δ, to thickness t.

Experiments on rectangular plates, Figure 10.8a, have shown that parts of the plate closer to the edges, Figure 10.8b, carry significantly more load than at the central portion of the plate. Various investigators developed different expressions for the effective width when ultimate loads are used. Marguerre's (1937) equation is based on the terminology of Figure 10.8c and is given by

$$b_e = b \left(\frac{\sigma_{cr}}{\sigma_u} \right)^{1/3}$$

(10.25)

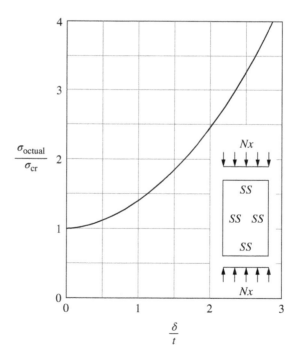

Figure 10.7   Stress versus deflection of a rectangular plate

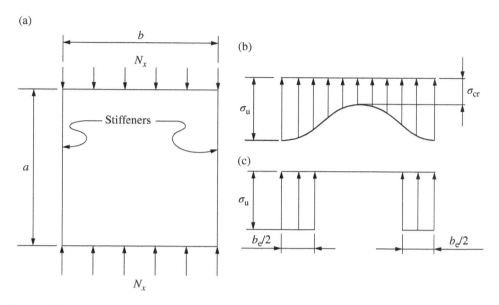

Figure 10.8 Stress distribution across plate: (a) rectangular plate, (b) elastic stress distribution, and (c) plastic stress distribution

while Koiter (1943) had a different expression in the form of

$$b_e = b\left[1.2\left(\frac{\sigma_{cr}}{\sigma_u}\right)^{0.4} - 0.65\left(\frac{\sigma_{cr}}{\sigma_u}\right)^{0.8} + 0.45\left(\frac{\sigma_{cr}}{\sigma_u}\right)^{1.2}\right]. \tag{10.26}$$

The ultimate load, P_u, carried by the plate can then be approximated by

$$P_u = \sigma_u b_e t. \tag{10.27}$$

10.3   Rectangular Plates with Various Boundary Conditions

Figure 10.9 shows a plate simply supported on sides $x = 0$ and $x = a$ and subjected to force N_x. The differential Eq. (10.18) becomes

$$\nabla^4 w = -\frac{N_x}{D}\frac{\partial^2 w}{\partial x^2}. \tag{10.28}$$

Let the solution be of the form

$$w = \sum_{m=1}^{\infty} f(y)\sin\frac{m\pi x}{a}. \tag{10.29}$$

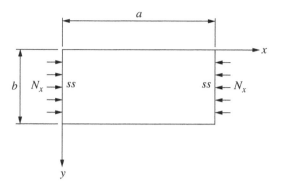

Figure 10.9 Rectangular plate with two sides simply supported

This solution satisfies the two boundary conditions $w = M_x = 0$ at $x = 0$ and $x = a$.
Substituting Eq. (10.29) into Eq. (10.28) results in

$$\frac{d^4f}{dy^4} - A\frac{d^2f}{dy^2} + Bf = 0 \tag{10.30}$$

where

$$A = \frac{2m^2\pi^2}{a^2}$$

$$B = \frac{m^4\pi^4}{a^4} - \frac{N_x}{D}\frac{m^2\pi^2}{a^2}.$$

The solution of Eq. (10.30) is

$$f(y) = C_1 e^{-\alpha y} + C_2 e^{\alpha y} + C_3 \cos\beta y + C_4 \sin\beta y \tag{10.31}$$

where

$$\alpha = \sqrt{\frac{m^2\pi^2}{a^2} + \sqrt{\frac{N_x}{D}\frac{m^2\pi^2}{a^2}}}$$

$$\beta = \sqrt{-\frac{m^2\pi^2}{a^2} + \sqrt{\frac{N_x}{D}\frac{m^2\pi^2}{a^2}}}.$$

Values of the constants C_1 through C_4 are obtained from the boundary conditions $y = 0$ and $y = b$.

Case 1

Side $y = 0$ is fixed and side $y = b$ is free. The four boundary conditions are for $y = 0$, deflection $w = 0$

$$\text{Rotation } \frac{\partial w}{\partial y} = 0$$

$$\text{for } y = b; \text{moment } M_y = 0 = \frac{\partial^2 w}{\partial y^2} + \mu \frac{\partial^2 w}{\partial x^2}$$

$$\text{Shear } Q = 0 = \frac{\partial^3 w}{\partial y^3} + (2 - \mu) \frac{\partial^3 w}{\partial x \partial y^2}.$$

From the first boundary condition, we get

$$C_1 + C_2 + C_3 = 0$$

and from the second boundary condition, we get

$$-\alpha C_1 + \alpha C_2 + \beta C_3 = 0$$

or

$$C_1 = -\frac{C_3}{2} + \frac{\beta C_4}{2\alpha}$$

and

$$C_2 = -\frac{C_3}{2} - \frac{\beta C_4}{2\alpha}.$$

Substituting C_1 and C_2 into Eq. (10.31) gives

$$f(y) = C_3 (\cos \beta y - \cosh \alpha y) + C_4 \left(\sin \beta y - \frac{\beta}{\alpha} \sinh \alpha y \right).$$

With this expression and Eq. (10.29), we can solve the last two boundary conditions. This results in two simultaneous equations. The critical value of the compressive force, N_x, is determined by equating the determinant of these equations to zero. This gives

$$2gh \left(g^2 + h^2 \right) \cos \beta b \cosh \alpha b = \frac{1}{\alpha \beta} \left(\alpha^2 h^2 - \beta^2 g^2 \right) \sin \beta b \sinh \alpha b \qquad (10.32)$$

where

$$g = \alpha^2 - \mu \frac{m^2 \pi^2}{a^2}$$

and

$$h = \beta^2 + \mu \frac{m^2 \pi^2}{a^2}.$$

For $m = 1$, the minimum value of Eq. (10.32) can be expressed in terms of stress as

$$\sigma_{cr} = \frac{\pi^2 E}{12(1-\mu^2)(b/t)^2} K \qquad (10.33)$$

where for $\mu = 0.25$,

$$K_{min} = 1.328. \qquad (10.34)$$

Case 2

Side $y = 0$ is simply supported and side $y = b$ is free. Again starting with Eq. (10.27) and satisfying the boundary conditions at $y = 0$ and $y = b$, we get a solution (Timoshenko and Gere 1961) identical to Eq. (10.33) with K given by

$$K = 0.456 + \frac{b^2}{a^2} \quad \text{for} \quad \mu = 0.25. \qquad (10.35)$$

Case 3

Sides $y = 0$ and $y = b$ are fixed. In this case, the minimum value of K in Eq. (10.33) is

$$K = 7.0 \quad \text{for} \quad \mu = 0.25. \qquad (10.36)$$

Example 10.3
Let the plate in Figure 10.5 be simply supported at $x = 0$ and $x = a$, simply supported at $y = 0$, and free at $y = b$. Calculate the required thickness if $a = 22$ inches, $b = 17$ inches, $N_x = 300$ 1b/inch, F.S. $= 2.0$, $\mu = 0.25$, $\sigma_y = 36$ ksi, and $E = 29,000$ ksi.

Solution
Assume $t = 0.25$ inch. From Eq. (10.35) the minimum value of $K = 0.456$ and Eq. (10.33) becomes

$$\sigma_{cr} = \frac{\pi^2 (29,000,000)}{12(0.9375)(17/0.25)^2} 0.456$$

$$= 2510 \, \text{psi which is less than the yield stress.}$$

Allowable stress $= 2510/2 = 1255$ psi. Actual stress $= 300/0.25 = 1200$ psi, which is acceptable.

10.4 Finite Difference Equations for Buckling

The finite difference method (Harrenstien and Alsmeyer 1959) is a powerful numerical tool for determining the approximate buckling of plates with irregular shapes, complicated boundary conditions, and nonuniform applied loads. It also serves as a handy tool for a obtaining a quick answer prior to performing more sophisticated, but lengthy, analysis. The basic differential equation of the buckling of a rectangular plate as given by Eq. (10.18) is

$$\frac{\partial^4 w}{\partial x^4} + 2\frac{\partial^4 w}{\partial x^2 \partial y^2} + \frac{\partial^4 w}{\partial y^4} = \frac{1}{D}\left(p + N_x\frac{\partial^2 w}{\partial x^2} + N_y\frac{\partial^2 w}{\partial y^2} + N_{xy}\frac{\partial^2 w}{\partial x \partial y}\right) \tag{10.37}$$

where N_x, N_y, and N_{xy} are entered as negative values. This equation can be transformed to a finite difference form (Jawad 2004) as

$$\frac{1}{\lambda^4}[20w_o - 8(w_E + w_W + w_N + w_S) + 2(w_{NE} + w_{SE} + w_{SW} + w_{NW})$$

$$+ (w_{NN} + w_{SS} + w_{EE} + w_{WW})] = -\left(\frac{1}{D\lambda^2}\right)[(w_E - 2w_o + w_W)N_x$$

$$+ (w_S - 2w_o + w_N)N_y + \left(\frac{1}{2}\right)(w_{NE} - w_{SE} - w_{NW} + w_{SW})N_{xy}]. \tag{10.38}$$

The following example illustrates the application of this equation in determining the buckling strength of a plate.

Example 10.4
Find the buckling strength of a simply supported square plate, Figure 10.10a, due to applied in-plane force $-N_x$. Let $N_y = N_{xy} = 0$. Assume (a) $\lambda = 10$ inches and (b) $\lambda = 5$ inches.

Solution
a. The grid pattern for $\lambda = 10$ inches is shown in Figure 10.10b. By inspection, $m = n = 1$. Applying Eq. (10.38) at node o gives

$$\left(\frac{1}{\lambda}\right)^4 [20 w_o - 8(0) + 2(0) + (-w_o - w_o - w_o - w_o)]$$

$$= \left(\frac{1}{D}\right)\left(\frac{1}{\lambda^2}\right)(-2w_o)(-N_x)$$

or

$$N_x = 8D/\lambda^2.$$

Substituting $\lambda = a/2$ in the previous expression gives the critical buckling value

$$(N_x)_{cr} = 32D/a^2.$$

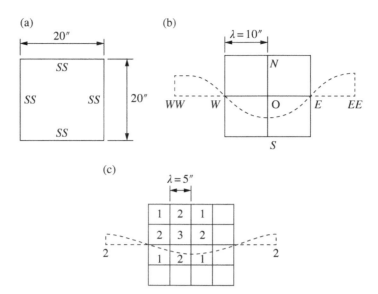

Figure 10.10 Finite difference layout of a square plate: (a) simply supported plate, (b) deflection of middle surface, and (c) finite difference layout

The theoretical critical value is given by

$(N_x)_{cr} = 4\pi^2 D/a^2$, which is about 19% larger than the calculated value.

b. The grid pattern for $\lambda = 5$ inches is shown in Figure 10.10c. Applying Eq. (10.38) at nodes 3, 2, and 1 gives

$$\left(\frac{1}{\lambda}\right)^4 [20w_3 - 8(4w_2) + 2(4w_1) + (0)] = \left(\frac{1}{D}\right)\left(\frac{1}{\lambda^2}\right)(2w_2 - 2w_3)(-N_x)$$

$$\left(\frac{1}{\lambda}\right)^4 [20w_2 - 8(w_3 + 2w_1) + 2(2w_2) + (w_2 - w_2)] = \left(\frac{1}{D}\right)\left(\frac{1}{\lambda^2}\right)(w_3 - 2w_2)(-N_x)$$

$$\left(\frac{1}{\lambda}\right)^4 [20w_1 - 8(2w_2) + 2(w_3) + (w_1 + w_1 - w_1 - w_1)] = \left(\frac{1}{D}\right)\left(\frac{1}{\lambda^2}\right)(w_2 - 2w_1)(-N_x).$$

These three equations can be written as

$$\begin{bmatrix} 20-2K & -32+2K & 8 \\ -8+K & 24-2K & -16 \\ 2 & -16+K & 20-2K \end{bmatrix} \begin{bmatrix} w_3 \\ w_2 \\ w_1 \end{bmatrix} = \begin{bmatrix} 0 \\ 0 \\ 0 \end{bmatrix}$$

where

$$K = \lambda^2 N_x/D.$$

A nontrivial solution of these three equations is obtained by setting the determinant to zero. This gives

$$K^3 - 24K^2 + 160K - 256 = 0.$$

The roots of this cubic equation are $K = 13.6569$, 8.000, and -2.3431. Taking the smallest root and substituting it in the expression for K gives

$$(N_x)_{cr} = 2.3431 \, D/\lambda^2$$

or

$$(N_x)_{cr} = 37.49 \, D/a^2.$$

This value is more accurate than that obtained in part (a) and is about 5% smaller than the theoretical value.

For rectangular plates with boundary conditions other than simply supported as well as other rectangular plates, a finer mesh is usually needed in order to improve the accuracy. Also, various values of m and n must be investigated in order to obtain minimum buckling load.

10.5   Other Aspects of Buckling

Temperature gradients are often encountered in many plate structures such as boiler casings and aircraft wings. The plates in these structures, Figure 10.11, are assumed to be continuous in the x-direction. They are also assumed to have no thermal gradients in that direction. The plates are also assumed to be supported by stiffeners at $y = 0$ and $y = b$. The stiffeners act as a heat sink resulting in a temperature gradient that varies in the y-direction in accordance with the equation

$$T = T_0 - T_1 \cos\left(\frac{2\pi y}{b}\right) \tag{10.39}$$

where T_0, and T_1 are constants that define the temperature variation along the y-axis. It is assumed by Boley and Weiner (1997) that the actual stress in the y-direction is zero and the actual stress in the x-direction is given by

$$\sigma_x = \alpha E T_1 \cos\left(\frac{2\pi y}{b}\right) \tag{10.40}$$

where $E =$ modulus of elasticity and $\alpha =$ coefficient of thermal expansion.

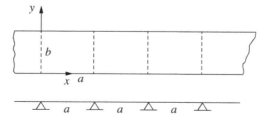

Figure 10.11   Continuous plate

Substituting Eq. (10.40) into Eq. (10.37) and solving the resultant equation for buckling result in

$$\sigma_{cr} = k \frac{\pi^2 E}{12(1-\mu^2)} \left(\frac{t}{b}\right)^2 \tag{10.41}$$

where

$$k = \frac{1}{2}(M_1)^2 \left\{ \left[(M^2+9)^4 + 4(M^2+1)^2 (M^2+9)^2 \right]^{1/2} - (M^2+9)^2 \right\} \tag{10.42}$$

and

$M = mb/a$

$M_1 = a/mb$

$m = 1, 2, \ldots$

Equation (10.42) is shown in Figure 10.12. A minimum value of k can be taken as $k = 3.848$. The critical temperature T_{1cr} is defined as

$$\frac{\alpha E}{2} T_{1cr} = \sigma_{cr}. \tag{10.43}$$

Substituting this expression into Eq. (10.41) gives

$$T_{1cr} = k \frac{\pi^2}{6\alpha(1-\mu^2)} \left(\frac{t}{b}\right)^2. \tag{10.44}$$

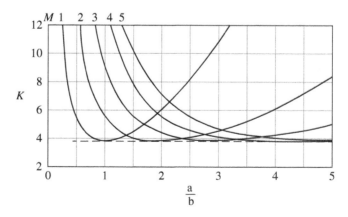

Figure 10.12   Plot of Eq. (10.42)

Example 10.5

A rectangular plate with lengths $a = 60$ inches and $b = 20$ inches has a temperature distribution in accordance with Eq. (10.39). It has a temperature of 800°F at $y = 0$ and $y = 20$ inches. The temperature at $y = 10$ inches is 1000°F. Let $E = 30 \times 10^6$ ksi, $\alpha = 6.5 \times 10^{-6}$ inch/inch/°F, yield stress = 30 ksi, and $\mu = 0.30$. What is the required thickness due to this temperature distribution assuming a F.S. of 1.3?

Solution

From Eq. (10.39) at $y = 0$,

$$800 = T_o - T_1 \tag{1}$$

and at $y = b/2$,

$$1000 = T_o + T_1. \tag{2}$$

Solving Eqs. (1) and (2) gives $T_o = 900°F$ and $T_1 = 100°F$. The actual maximum thermal stress in the plate is obtained from Eq. (10.40) as

$$\sigma = \left(6.5 \times 10^{-6}\right)\left(30 \times 10^6\right)(100)$$

$$= 19,500 \, \text{psi.}$$

The critical buckling stress can be taken as this number multiplied by the F.S. of 1.3.

$$\sigma_{cr} = 19,500 \times 1.3 = 25,350 \, \text{psi.}$$

Since this value is lower than the yield stress of the material, buckling controls. Substituting this value into the buckling Eq. (10.41) gives

$$25,350 = (3.848)\frac{\pi^2 \left(30 \times 10^6\right)}{12\left(1 - 0.3^2\right)}\left(\frac{t}{20}\right)^2.$$

Solving for thickness results in

$$t = 0.31 \, \text{inch.}$$

10.6   Application of Buckling Expressions to Design Problems

Many codes utilize the expressions of Section 10.3 to establish buckling criteria for various members. The American Institute of Steel Construction Manual (AISC 2013) assumes the buckling stress of unsupported members in compression not to exceed the yield stress of the material. Thus, Eq. (10.33) can be written as

$$\sigma_y = \frac{\pi^2 E}{12\left(1 - \mu^2\right)(b/t)^2}K \tag{10.45}$$

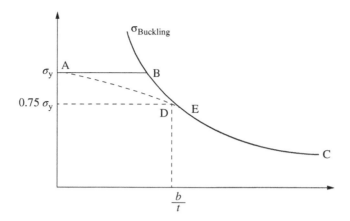

Figure 10.13   Elastic–plastic buckling

or for steel members with $\mu = 0.3$,

$$\frac{b}{t} = 0.903\sqrt{\frac{KE}{\sigma_y}}. \tag{10.46}$$

Equation 10.45 is based on the assumption that the interaction between the buckling stress and the yield stress in designated by points ABC in Figure 10.13. However, due to residual stress in structural members due to forming, the actual interaction curve is given by points ADC, and Eq. (10.46) is modified by a factor of 0.7 as

$$\frac{b}{t} = 0.633\sqrt{\frac{KE}{\sigma_y}}. \tag{10.47}$$

10.6.1   Single- and Double-Angle Struts

Leg b of the angle shown in Figure 10.14a and b is assumed free at point B. Point A is assumed simply supported because it can rotate due to deflections. Thus, Eq. (10.35) is applicable with a minimum value of $K = 0.456$ and Eq. (10.46) becomes

$$\frac{b}{t} = 0.45\sqrt{\frac{E}{\sigma_y}}.$$

10.6.2   Stems of Ts

For this case, point A in Figure 10.14c is assumed fixed due to the much thicker flange, and point B is taken as free. The K value is taken from Eq. (10.34) and expression 10.47 gives

$$\frac{b}{t} = 0.78\sqrt{\frac{E}{\sigma_y}}.$$

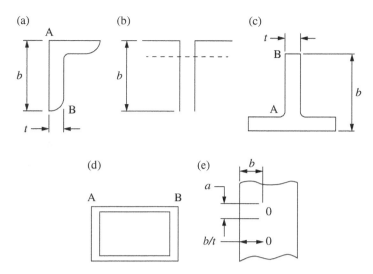

Figure 10.14 Various structural members: (a) structural angle, (b) double structural angles, (c) structural *T*, (d) box section, and (e) perforated plate

The AISC reduces this value further down to

$$\frac{b}{t} = 0.75\sqrt{\frac{E}{\sigma_y}}.$$

10.6.3 *Flanges of Box Sections*

Points A and B in Figure 10.14d are conservatively taken as simply supported. In this case, $k = 4.0$ and Eq. (10.47) becomes

$$\frac{b}{t} = 1.34\sqrt{\frac{E}{\sigma_y}}.$$

The AISC increases this value to match experimental data and it becomes

$$\frac{b}{t} = 1.40\sqrt{\frac{E}{\sigma_y}}.$$

10.6.4 *Perforated Cover Plates*

In Figure 10.14e, the plate between the perforation and edge is assumed fixed because additional rigidity is obtained from the continuous areas between the perforations. The dimension of the perforated plate is assumed to be $a/b = 1.0$. This ratio results in a K value of about 7.69. This is higher than that given by Eq. (10.36), which is based on the smallest possible value.

Equation 10.47 becomes

$$\frac{b}{t} = 1.86\sqrt{\frac{E}{\sigma_y}}.$$

10.6.5 Other Compressed Members

Other members are assumed to have a K value that varies between 4.0 for simply supported edges and 7.0 for fixed edges. The AISC uses a value of $K = 4.90$. This gives

$$\frac{b}{t} = 1.49\sqrt{\frac{E}{\sigma_y}}.$$

Another standard that uses the plate buckling equations to set a criterion is the American Association of State Highway and Transportation Officials (AASHTO 1996). The equations are very similar to those of AISC.

A theoretical solution of the buckling of rectangular plates due to various loading and boundary conditions is available in numerous references. Two such references are Timoshenko and Gere (1961) and Iyengar (1988). Timoshenko discusses mainly isotropic plates, whereas Iyengar handles composite plates.

Various NASA publications are also available for the solution of the buckling of rectangular plates with various loading and boundary conditions. NASA's *Handbook of Structural Stability* consists of five parts and contains numerous theoretical background and design aids. Part I, edited by Gerard and Becker, includes various classical buckling solutions for flat plates. Part II, edited by Becker, is for buckling of composite elements. Part III is for buckling of curved plates and shells, and Part IV discusses failure of plates and composite elements. Parts III and IV were edited by Gerard and Becker. Compressive strength of flat stiffened panels is given in Part V, which is edited by Gerard.

Problems

10.1 What is the required thickness of a fixed circular plate subjected to a lateral pressure of 3 psi and in-plane compressive force of 500 lbs/inch if $a = 40$ inches, $\mu = 0.30$, $E = 25,000$ ksi, allowable stress in bending $= 25,000$ psi, and F.S. for buckling $= 4.0$?

10.2 What is the effect of N_r on the bending moments in a circular plate if it were in tension rather than compression?

10.3 What is the required thickness of the plate in Figure 10.5 if $a = 40$ inches, $b = 15$ inches, $N_x = 400$ lbs/inch, $E = 16,000$ ksi, and $\mu = 0.33$? Use a F.S. of 4.0. Use increments of 1/16 inch for thickness.

10.4 Determine the actual expression of K in Eq. (10.34). Plot K versus a/b for m values of 1, 2, and 3 and a/b values from 1.0 to 5.0.

11

Finite Element Analysis

11.1 Definitions

The finite element method is a powerful numerical tool for calculating stress in complicated shell and plate structures that are difficult to analyze by classical plate and shell theories. The method consists of subdividing a given domain into small elements connected at the nodal points as shown in Figure 11.1. The mathematical formulation consists of combining the governing equations of each of the elements to form a solution for the domain that satisfies the boundary conditions. The approximations associated with finite element solutions depend on many variables such as the type of element selected, the number of elements used to model the domain, and the boundary conditions.

The complete derivation of the various equations for one-, two-, and three-dimensional elements is beyond the scope of this book. However, a few equations are derived here to demonstrate the basic concept of finite element formulation and its applicability to the solution of plates and shells.

We begin the derivations by defining various elements (Figure 11.2) and terms. Figure 11.2a shows a one-dimensional element in the x-direction with two nodal points, i and j. Figure 11.2b shows a two-dimensional triangular element in the x, y plane with nodal points i, j, and k. And Figure 11.2c shows a three-dimensional rectangular brick element with eight nodal points.

Let the matrix $[\delta]$ define the displacements within an element. The size of the displacement matrix (Weaver and Johnston 1984) depends on the complexity of the element being considered. The matrix $[q]$ defines nodal point displacements of an element, while matrix $[F]$ defines the applied loads at the nodal points. The size of matrices $[q]$ and $[F]$ depends on

Stress in ASME Pressure Vessels, Boilers, and Nuclear Components, First Edition. Maan H. Jawad.
© 2018, The American Society of Mechanical Engineers (ASME), 2 Park Avenue,
New York, NY, 10016, USA (www.asme.org). Published 2018 by John Wiley & Sons, Inc.

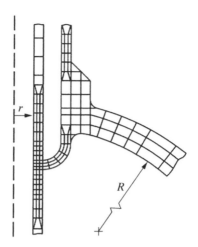

Figure 11.1   Nozzle-to-head attachment

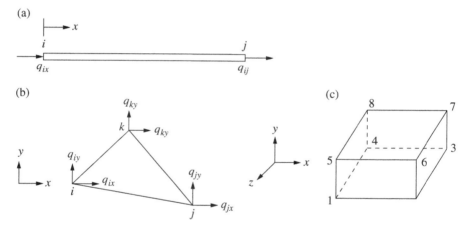

Figure 11.2   One-, two-, and three-dimensional elements: (a) rod element, (b) triangular element, and (c) solid element

the type and geometry of the element being considered. For the one-dimensional element in Figure 11.2a, the quantities $[q]$ and $[F]$ are defined as

$$[q] = \begin{bmatrix} q_{ix} \\ q_{jx} \end{bmatrix}$$
$$[F] = \begin{bmatrix} F_{ix} \\ F_{jx} \end{bmatrix}$$

(11.1)

Similarly, the $[q]$ and $[F]$ matrices for the two-dimensional triangular element in Figure 11.2b are expressed by

$$[q] = \begin{bmatrix} q_{ix} \\ q_{iy} \\ q_{jx} \\ q_{jy} \\ q_{kx} \\ q_{ky} \end{bmatrix} \quad [F] = \begin{bmatrix} F_{ix} \\ F_{iy} \\ F_{jx} \\ F_{jy} \\ F_{kx} \\ F_{ky} \end{bmatrix}. \tag{11.2}$$

The shape function matrix [N] defines the relationship between a function at the nodal points and the same function within the element. Thus, the relationship between the nodal deflection [q] in Figure 11.2a and the general deflection δ at any point in the one-dimensional element is expressed as

$$\delta = \begin{bmatrix} N_i & N_j \end{bmatrix} \begin{bmatrix} q_i \\ q_j \end{bmatrix}$$

$$= [N][q]$$

while the relationship between the nodal displacements [q] and the general displacements [δ] in Figure 11.2b for a two-dimensional element is

$$\begin{bmatrix} u \\ v \end{bmatrix} = \begin{bmatrix} N_i & 0 & N_j & 0 & N_k & 0 \\ 0 & N_i & 0 & N_j & 0 & N_k \end{bmatrix} \begin{bmatrix} q_{ix} \\ q_{iy} \\ q_{jx} \\ q_{jy} \\ q_{kx} \\ q_{ky} \end{bmatrix}$$

or

$$[\delta] = [N][q]. \tag{11.3}$$

Let the strain–displacement matrix [d] define the relationship between the strains in a continuum to the displacements in accordance with

$$[\varepsilon] = [d][\delta]. \tag{11.4}$$

Strain in an element can also be expressed in terms of the deflection of the nodal points. Substituting Eq. (11.3) into Eq. (11.4) gives

$$[\varepsilon] = [d][N][q]. \tag{11.5}$$

Equation (11.5) can also be written as

$$[\varepsilon] = [B][q] \tag{11.6}$$

where

$$[B] = [d][N].$$ (11.7)

The stress–strain relationship is obtained as

$$[\sigma] = [D][\varepsilon] - [D][\varepsilon_o]$$
$$= [D][B][q] - [D][\varepsilon_o]$$ (11.8)

where $[\varepsilon_o]$ is the initial strain in a domain and $[\varepsilon]$ is the total strain.

With these definitions, the basic finite element equations can now be derived. The strain energy for a differential element of volume dV is

$$dU = \frac{1}{2}[\varepsilon]^T[\sigma] - \frac{1}{2}[\varepsilon_o]^T[\sigma].$$ (11.9)

The total strain energy is given by

$$U = \int_V \frac{1}{2}\left([\varepsilon]^T[\sigma] - [\varepsilon_o]^T[\sigma]\right)dV.$$ (11.10)

Substituting Eq. (11.8) into Eq. (11.10) yields

$$U = \frac{1}{2}\{[q]^T[B]^T[D][B][q] - 2[q]^T[B]^T[D][\varepsilon_o] + [\varepsilon_o]^T[D][\varepsilon_o]\}dV.$$ (11.11)

The external work due to the nodal loads $[F]$ is

$$W_F = [F]^T[q].$$ (11.12)

The external work due to surface pressure, $[p]$, is

$$W_p = ([u][p])\ ds$$

or

$$W_p = \int_s \left([q]^T[N]^T[p]\right)\ ds.$$ (11.13)

The potential energy of one element is

$$\Pi = U - \left(W_F + W_p\right)$$

or for the whole system

$$\Pi = \sum_{e=1}^{E}\left[U^e - \left(W_F^e + W_p^e\right)\right]$$ (11.14)

where e refers to any given element.

The minimum potential energy is obtained from

$$\frac{\partial \Pi}{\partial q} = 0$$

or

$$\frac{\partial \Pi}{\partial q} = \sum_{e=1}^{E} \left[\int_V [B_e]^T [D_e][B_e] dV [q] \right.$$

$$- \int_V [B_e]^T [D_e][\varepsilon_0] dV \qquad (11.15)$$

$$\left. - \int_S [N_e]^T [p_e] ds \right] - F_e = 0.$$

The quantity

$$[B_e]^T [D_e][B_e] dV$$

is called the stiffness matrix of an element and is written as

$$[K_e] = \int_V [B_e]^T [D_e][B_e] dV. \qquad (11.16)$$

Hence, the finite element equation becomes

$$\sum_{e=1}^{E} [K_e][q] = \sum_{e=1}^{E} \left[\int_V [B_e]^T [D_e][\varepsilon_0] dV \right.$$

$$\left. + \int_S [N_e]^T [p_e] ds \right] + F_e \qquad (11.17)$$

which can be abbreviated as

$$[K_e][q] = [F] \qquad (11.18)$$

where $[F]$ = applied forces.

Equation (11.17) is the basic finite element equation for a domain.

11.2   One-Dimensional Elements

In formulating the finite element equations, the shape of the element and other functions such as applied loads, deflections, strains, and stresses are approximated by a polynomial. The size of the polynomial depends on the degrees of freedom at the nodal points and the accuracy required. Hence, for the one-dimensional element shown in Figure 11.2a, a polynomial for a function such as deflection (Grandin 1986) may be expressed as

$$\delta = C_1 + C_2 x \tag{11.19}$$

where x = length along the x-axis and C_1 and C_2 are constants.

The polynomial given by Eq. (11.19) can be written as a function of two matrices $[g]$ and $[C]$ as

$$\delta = [g][C] \tag{11.20}$$

$$\delta = [1 \ \ x] \begin{bmatrix} C_1 \\ C_2 \end{bmatrix}.$$

At the nodal points x_i and x_j, Eq. (11.20) becomes

$$\begin{bmatrix} q_i \\ q_j \end{bmatrix} = \begin{bmatrix} 1 & x_i \\ 1 & x_j \end{bmatrix} \begin{bmatrix} C_1 \\ C_2 \end{bmatrix}. \tag{11.21}$$

Define $[h]$ as the relationship between $[C]$ and $[q]$ at the nodal points. Thus, Eq. (11.21) becomes

$$\begin{bmatrix} q_i \\ q_j \end{bmatrix} = [h][C]$$

where

$$[h] = \begin{bmatrix} 1 & x_i \\ 1 & x_j \end{bmatrix}.$$

Solving for the matrix $[C]$ gives

$$\begin{bmatrix} C_1 \\ C_2 \end{bmatrix} = [h]^{-1} \begin{bmatrix} q_i \\ q_j \end{bmatrix}$$

$$\begin{bmatrix} C_1 \\ C_2 \end{bmatrix} = \frac{1}{x_i - x_j} \begin{bmatrix} x_j & -x_i \\ -1 & 1 \end{bmatrix} \begin{bmatrix} q_i \\ q_j \end{bmatrix} \tag{11.22}$$

or, from Eq. (11.20), the function at any point is

$$\delta = [g][h]^{-1} \begin{bmatrix} q_i \\ q_j \end{bmatrix}$$

or

$$\delta = \frac{1}{x_j - x_i} \left[(x_j - x) \quad (-x_i + x) \right] \begin{bmatrix} q_i \\ q_j \end{bmatrix}. \tag{11.23}$$

The quantity $[g][h]^{-1}$ relates the deflection at the nodal points to that within the element. It is called the shape function and is designated as $[N]$. Thus,

$$[N] = [g][h]^{-1}. \tag{11.24}$$

Then Eq. (11.23) can be written as

$$\delta = [N] \begin{bmatrix} q_i \\ q_j \end{bmatrix} \tag{11.25}$$

where

$$[N] = \frac{1}{x_j - x_i} \left[(x_j - x) \quad (-x_i + x) \right]. \tag{11.26}$$

Once the shape function for a linear element is established, the governing stiffness expression, Eq. (11.18), can also be determined. Thus, for the one-dimensional element shown in Figure 11.2a, the general deflection function is expressed by Eq. (11.19), and the shape function $[N]$ is given by Eq. (11.26). From Hooke's law, the strain in an axial member is expressed as

$$\varepsilon = \frac{d}{dx} u$$

and from Eqs. (11.5) and (11.6)

$$[\varepsilon] = \frac{d}{dx} [N][q]$$

or

$$[\varepsilon] = [B] \begin{bmatrix} q_i \\ q_j \end{bmatrix} \tag{11.27}$$

where

$$[B] = \frac{[-1 \quad 1]}{x_j - x_i} = \frac{1}{L} [-1 \quad 1].$$

For a uniaxial body without initial strain,

$$[\sigma] = [E][\varepsilon].$$

Hence,

$$[D] = [E].$$

From Eq. (11.16) the value of the stiffness matrix $[K]$ becomes

$$[K] = \frac{AE}{L} \begin{bmatrix} 1 & -1 \\ -1 & 1 \end{bmatrix}. \tag{11.28}$$

From Eq. (11.17), the first term on the right-hand side is due to the thermal effect and reduces to

$$\alpha EA(\Delta T) \begin{bmatrix} -1 \\ 1 \end{bmatrix}. \tag{11.29}$$

The second term on the right-hand side of Eq. (11.17) is for the surface loads. In this case, the surface loads can be applied only at the nodal points i and j. Hence, when p_x is applied at node i,

$$\int_S [N]^T [p_x] ds = p_x \begin{bmatrix} 1 \\ 0 \end{bmatrix} \int ds = p_x A_i \begin{bmatrix} 1 \\ 0 \end{bmatrix}. \tag{11.30}$$

When p_x is applied at node j,

$$\int_S [N]^T [p_x] ds = p_x A_i \begin{bmatrix} 1 \\ 0 \end{bmatrix}. \tag{11.31}$$

The complete finite element equation for one-dimensional elements is obtained by combining Eqs. (11.28) through (11.31)

$$\frac{AE}{L} \begin{bmatrix} 1 & -1 \\ -1 & 1 \end{bmatrix} \begin{bmatrix} q_i \\ q_j \end{bmatrix} = \alpha EA(\Delta T) \begin{bmatrix} -1 \\ 1 \end{bmatrix} + A_i p_x \begin{bmatrix} 1 \\ 0 \end{bmatrix} + A_j p_x \begin{bmatrix} 1 \\ 0 \end{bmatrix} + F$$

or

$$[K][q] = [F]. \tag{11.32}$$

Stress in the member is obtained from Eq. (11.8) as

$$\sigma = [E][B][q] = \frac{E}{L}[-1 \quad 1] \begin{bmatrix} q_i \\ q_j \end{bmatrix} - E \propto (\Delta T). \tag{11.33}$$

Example 11.1

Find the stress in the tapered conical shell, Figure 11.3. The shell is subjected to a force of 50 kips at point A. The cross-sectional area of the cone increases from 6 inch^2 at the right end to 15 inch^2 at the left end. The shell is also subjected to a uniform decrease in temperature of 50°F. Assume the shell to be subdivided into three equal lengths, and let the coefficient of thermal expansion be equal to 7×10^{-6} inch/inch/°F. Also, let the modulus of elasticity be equal to 30×10^6 psi and Poisson's ratio equal to 0.3.

Solution

Element 1   The stiffness matrix from Eq. (11.28) is

$$[K_1] = \frac{\bar{A}E}{L}\begin{bmatrix} 1 & -1 \\ -1 & 1 \end{bmatrix} = 10^7 \begin{bmatrix} 10.125 & -10.125 \\ -10.125 & 10.125 \end{bmatrix}.$$

The thermal force from Eq. (11.29) is

$$\alpha\bar{A}E(-\Delta T)\begin{bmatrix} -1 \\ 1 \end{bmatrix} = 10^3 \begin{bmatrix} 141.75 \\ -141.75 \end{bmatrix}.$$

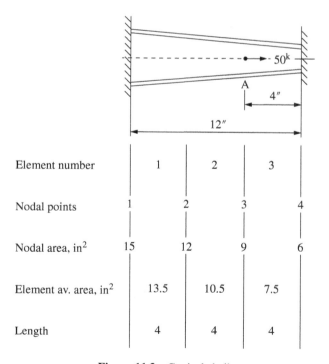

	Element number	1	2	3
Nodal points		1	2	3	4
Nodal area, in^2		15	12	9	6
Element av. area, in^2		13.5	10.5	7.5
Length		4	4	4

Figure 11.3   Conical shell

and Eq. (11.18) gives

$$10^7 \begin{bmatrix} 10.125 & -10.125 \\ -10.125 & 10.125 \end{bmatrix} \begin{bmatrix} q_1 \\ q_2 \end{bmatrix} = 10^3 \begin{bmatrix} 141.75 \\ -141.75 \end{bmatrix}.$$

Element 2   The governing Eq. (11.18) for element 2 is

$$10^7 \begin{bmatrix} 7.875 & -7.875 \\ -7.875 & 7.875 \end{bmatrix} \begin{bmatrix} q_2 \\ q_3 \end{bmatrix} = 10^3 \begin{bmatrix} 110.25 \\ -110.25 \end{bmatrix}.$$

Element 3

$$[K_3] = \frac{\bar{A}E}{L} \begin{bmatrix} 1 & -1 \\ -1 & 1 \end{bmatrix} = 10^7 \begin{bmatrix} 5.625 & -5.625 \\ -5.625 & 5.625 \end{bmatrix}.$$

The thermal force is

$$\alpha \bar{A} E(-\Delta T) \begin{bmatrix} -1 \\ 1 \end{bmatrix} = 10^3 \begin{bmatrix} 78.75 \\ -78.75 \end{bmatrix}.$$

The nodal forces are

$$= 10^3 \begin{bmatrix} 50 \\ 0 \end{bmatrix}$$

and Eq. (11.18) becomes

$$10^7 \begin{bmatrix} 5.625 & -5.625 \\ -5.625 & 5.625 \end{bmatrix} \begin{bmatrix} q_3 \\ q_4 \end{bmatrix} = 10^3 \begin{bmatrix} 128.75 \\ -78.75 \end{bmatrix}.$$

Combining the matrices for elements 1, 2, and 3 (Table 11.1) gives

$$10^7 \begin{bmatrix} 10.125 & -10.125 & 0 & 0 \\ -10.125 & 18.000 & -7.875 & 0 \\ 0 & -7.875 & 13.500 & -5.625 \\ 0 & 0 & -5.625 & 5.625 \end{bmatrix} \begin{bmatrix} q_1 \\ q_2 \\ q_3 \\ q_4 \end{bmatrix} = 10^3 \begin{bmatrix} 141.75 \\ -31.50 \\ 18.50 \\ -78.75 \end{bmatrix}.$$

Because the deflections q_1 and q_4 are zero at the supports, we can delete the first and last rows and columns from the stiffness matrix and the previous matrix reduces to

$$10^7 \begin{bmatrix} 18.000 & -7.875 \\ -7.875 & 13.500 \end{bmatrix} \begin{bmatrix} q_2 \\ q_3 \end{bmatrix} = 10^3 \begin{bmatrix} -31.50 \\ 18.50 \end{bmatrix}.$$

Table 11.1   Total stiffness and force matrices

Stiffness matrix [K]

F/q	1	2	3	4
1	10.125	−10.125		
2	−10.125	10.125		
		7.875	−7.875	
3		−7.875	7.875	
			5.625	−5.625
4			−5.625	5.625

Load matrix [F]

Node	Force
1	141.75
2	−141.75
	110.25
3	−110.25
	128.75
4	−78.75

Solving for the values of q_2 and q_3 results in

$$\begin{bmatrix} q_2 \\ q_3 \end{bmatrix} = \frac{1}{10^4} \begin{bmatrix} -1.545 \\ 0.469 \end{bmatrix}.$$

The strain expression in each element is given by Eq. (11.27) as

$$\varepsilon = \frac{1}{L}\left(-q_i + q_j\right)$$

$$\varepsilon_1 = -0.386 \times 10^{-4}$$

$$\varepsilon_2 = 0.504 \times 10^{-4}$$

$$\varepsilon_3 = -0.117 \times 10^{-4}.$$

The stress is obtained from Eq. (11.33) as

$$\sigma = E\varepsilon - \alpha E(-\Delta T)$$

$$= 30 \times 10^6 \varepsilon + 10,500$$

$$\begin{bmatrix} \sigma_1 \\ \sigma_2 \\ \sigma_3 \end{bmatrix} = \begin{bmatrix} 9340 \\ 12,010 \\ 10,150 \end{bmatrix}.$$

A classical theoretical solution of this simple problem can be obtained for comparison purposes. We can let the right end grow freely due to temperature and applied load. We then apply a load at the end to let the deflection at the right end be equal to zero. Using for the deflection due to loads the equation

$$w = \int \frac{F dx}{EA_x}$$

$$= \frac{F}{E} \int \frac{dx}{15 - (9x/L)}$$

we get $q_2 = -1.574$ and $q_3 = 0.449$

$$\begin{bmatrix} \sigma_1 \\ \sigma_2 \\ \sigma_3 \end{bmatrix} = \begin{bmatrix} 9290 \\ 11,930 \\ 10,040 \end{bmatrix}.$$

The calculated stress in Example 11.1 is different in each of the elements. This causes a discontinuity in stress at internal nodal points joining two elements. To overcome this, a procedure (Segerlind 1976), called the conjugate stress method, is used to average the stresses at the nodal points. It calculates an approximate average stress value at the nodal points from the following equation

$$[Q][\bar{\sigma}] = [R] \tag{11.34}$$

where
$[Q]$ = A function of the $[N]$ matrix
$[\bar{\sigma}]$ = Stress at the nodal points, called conformal stress
$[R]$ = A function of the element stress, called conjugate stress

In accordance with the theory of conjugate stress approximations, the matrices $[Q]$ and $[R]$ for an element are calculated from the quantities

$$[Q] = \int_V [N]^T [N] dV \tag{11.35}$$

and

$$[R] = \int_V [\sigma][N]^T dV \tag{11.36}$$

where $[\sigma]$ is the stress in the element.

In many applications the member axis, which is used to determine the stiffness and load matrices, does not coincide with the global axis of the structure as illustrated in Figure 11.4.

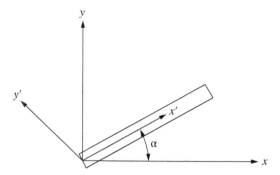

Figure 11.4 Element orientation

In order to accommodate this condition, the member orientation with respect to the global axes needs to be taken into consideration. The resulting stiffness matrix (Wang 1986) is of the form

$$[K_G] = \frac{AE}{L} \begin{bmatrix} k_1 & k_2 & -k_1 & -k_2 \\ & k_3 & -k_2 & -k_3 \\ & & k_1 & k_2 \\ \text{symmetric} & & & k_3 \end{bmatrix} \tag{11.37}$$

where
$k_1 = \cos^2\alpha$, $k_2 = \cos\alpha \sin\alpha$, $k_3 = \sin^2\alpha$
α = Angle shown in Figure 11.4 and measured counterclockwise from the positive x-axis, and load in each member is expressed as

$$F = \frac{EA}{L} \left(-q_{ix} \cos\alpha - q_{iy} \sin\alpha + q_{jx} \cos\alpha + q_{jy} \sin\alpha \right). \tag{11.38}$$

11.3 Linear Triangular Elements

From Figure 11.2b it is seen that each element has three nodal points, and each nodal point has two degrees of freedom. Hence the displacement within the element is expressed by the following polynomial

$$u = C_1 + C_2 x + C_3 y$$
$$v = C_4 + C_5 x + C_6 y \tag{11.39}$$

where u and v are the deflection in the x- and y-axes, respectively. In matrix form these equations are written as

$$\begin{bmatrix} u \\ v \end{bmatrix} = [g][C]$$

where

$$[g] = \begin{bmatrix} 1 & x & y & 0 & 0 & 0 \\ 0 & 0 & 0 & 1 & x & y \end{bmatrix} \qquad (11.40)$$

and

$$[C] = \begin{bmatrix} C_1 \\ \cdot \\ \cdot \\ \cdot \\ C_6 \end{bmatrix}.$$

The shape matrix $[N]$ is obtained from Eq. (11.24) and is expressed as

$$[N] = \begin{bmatrix} N_i & 0 & N_j & 0 & N_k & 0 \\ 0 & N_i & 0 & N_j & 0 & N_k \end{bmatrix} \qquad (11.41)$$

where N_i, N_j, and N_k are defined as

$$N_i = \frac{1}{2\Delta}(a_i + b_i x + c_i y)$$

$$N_j = \frac{1}{2\Delta}(a_j + b_j x + c_j y)$$

$$N_k = \frac{1}{2\Delta}(a_k + b_k x + c_k y)$$

$$a_i = x_j y_k - x_k y_j, \quad b_i = y_j - y_k, \quad c_i = x_k - x_j$$

$$a_j = x_k y_i - x_i y_k, \quad b_j = y_k - y_i, \quad c_j = x_i - x_k$$

$$a_k = x_i y_j - x_j y_i, \quad b_k = y_i - y_j, \quad c_k = x_j - x_i$$

$$\Delta = \frac{1}{2}\begin{vmatrix} 1 & x_i & y_i \\ 1 & x_j & y_j \\ 1 & x_k & y_k \end{vmatrix}$$

2Δ = Area of triangle with coordinates $x_i y_i$, $x_j y_j$, and $x_k y_k$.

The u and v expressions within the element are

$$\begin{bmatrix} u \\ v \end{bmatrix} = [N] \begin{bmatrix} q_{ix} \\ q_{iy} \\ q_{jx} \\ q_{jy} \\ q_{kx} \\ q_{ky} \end{bmatrix}.$$

The strain–deflection relationship is obtained from

$$
\begin{bmatrix} \varepsilon_x \\ \varepsilon_y \\ \gamma_{xy} \end{bmatrix} = \begin{bmatrix} \dfrac{\partial}{\partial x} & 0 \\ 0 & \dfrac{\partial}{\partial y} \\ \dfrac{\partial}{\partial y} & \dfrac{\partial}{\partial x} \end{bmatrix} \begin{bmatrix} u \\ v \end{bmatrix}. \tag{11.42}
$$

This strain expression can be designated as

$$
[\varepsilon] = [d]\begin{bmatrix} u \\ v \end{bmatrix} = [d][N][q]
$$

or

$$
[\varepsilon] = [B][q]
$$

where

$$
[B] = [d][N]
$$

$$
= \frac{1}{2\Delta}\begin{bmatrix} b_i & 0 & b_j & 0 & b_k & 0 \\ 0 & c_i & 0 & c_j & 0 & c_k \\ c_i & b_i & c_j & b_j & c_k & b_k \end{bmatrix}. \tag{11.43}
$$

The stress matrix $[\sigma]$ is calculated from Eq. (11.8) where $[D]$ is defined as follows:
For a plane stress formulation, the stress–strain matrix $[D]$ is given by

$$
[D] = \frac{E}{1-\mu^2}\begin{bmatrix} 1 & \mu & 0 \\ \mu & 1 & 0 \\ 0 & 0 & \dfrac{1-\mu}{2} \end{bmatrix}. \tag{11.44a}
$$

For a plane strain formulation,

$$
\varepsilon_z = \gamma_{yz} = \gamma_{xz} = 0
$$

$$
[D] = \frac{E(1-\mu)}{(1+\mu)(1-2\mu)}\begin{bmatrix} 1 & \dfrac{\mu}{1-\mu} & 0 \\ \dfrac{\mu}{1-\mu} & 1 & 0 \\ 0 & 0 & \dfrac{1-2\mu}{2(1-\mu)} \end{bmatrix}. \tag{11.44b}
$$

The stiffness matrix is calculated from Eq. (11.16). The result can be expressed as

$$K = \frac{Et}{4\Delta} \begin{bmatrix} k_{11} & k_{12} & k_{13} & k_{14} & k_{15} & k_{16} \\ & k_{22} & k_{23} & k_{24} & k_{25} & k_{26} \\ & & k_{33} & k_{34} & k_{35} & k_{36} \\ & & & k_{44} & k_{45} & k_{46} \\ & \text{symmetric} & & & k_{55} & k_{56} \\ & & & & & k_{66} \end{bmatrix} \tag{11.45}$$

where

$$k_{11} = k_1(y_2 - y_3)^2 + k_3(x_3 - x_2)^2$$

$$k_{12} = k_2(x_3 - x_1)(y_2 - y_3) + k_3(x_3 - x_2)(y_2 - y_3)$$

$$k_{13} = k_1(y_2 - y_3)(y_3 - y_1) + k_3(x_3 - x_2)(x_1 - x_3)$$

$$k_{14} = k_2(x_1 - x_3)(y_2 - y_3) + k_3(x_3 - x_2)(y_3 - y_1)$$

$$k_{15} = k_1(y_1 - y_2)(y_2 - y_3) + k_3(x_2 - x_1)(x_3 - x_2)$$

$$k_{16} = k_2(x_2 - x_1)(y_2 - y_3) + k_3(x_3 - x_2)(y_1 - y_2)$$

$$k_{22} = k_1(x_3 - x_2)^2 + k_3(y_2 - y_3)^2$$

$$k_{23} = k_2(x_3 - x_2)(y_3 - y_1) + k_3(x_1 - x_3)(y_2 - y_3)$$

$$k_{24} = k_1(x_3 - x_2)(x_1 - x_3) + k_3(y_2 - y_3)(y_3 - y_1)$$

$$k_{25} = k_2(x_3 - x_2)(y_1 - y_2) + k_3(x_2 - x_1)(y_2 - y_3)$$

$$k_{26} = k_1(x_2 - x_1)(x_3 - x_2) + k_3(y_1 - y_2)(y_2 - y_3)$$

$$k_{33} = k_1(y_3 - y_1)^2 + k_3(x_1 - x_3)^2$$

$$k_{34} = k_2(x_1 - x_3)(y_3 - y_1) + k_3(x_1 - x_3)(y_3 - y_1)$$

$$k_{35} = k_1(y_1 - y_2)(y_3 - y_1) + k_3(x_1 - x_3)(x_2 - x_1)$$

$$k_{36} = k_2(x_2 - x_1)(y_3 - y_1) + k_3(x_1 - x_3)(y_1 - y_2)$$

$$k_{44} = k_1(x_1 - x_3)^2 + k_3(y_3 - y_1)^2$$

$$k_{45} = k_2(x_1 - x_3)(y_1 - y_2) + k_3(x_2 - x_1)(y_3 - y_1)$$

$$k_{46} = k_1(x_1 - x_3)(x_2 - x_1) + k_3(y_1 - y_2)(y_3 - y_1)$$

$$k_{55} = k_1(y_1 - y_2)^2 + k_3(x_2 - x_1)^2$$

$$k_{56} = k_2(x_2 - x_1)(y_1 - y_2) + k_3(x_2 - x_1)(y_1 - y_2)$$

$$k_{66} = k_1(x_2 - x_1)^2 + k_3(y_1 - y_2)^2$$

and

For Plane Stress

$$k_1 = \frac{1}{1-\mu^2} \quad k_2 = \frac{\mu}{1-\mu^2} \quad k_3 = \frac{1}{2(1+\mu)}$$

For Plane Strain

$$k_1 = \frac{1-\mu}{(1+\mu)(1-2\mu)} \quad k_2 = \frac{\mu}{(1+\mu)(1-2\mu)}$$

$$k_3 = \frac{1}{2(1+\mu)}.$$

After assembling the stiffness matrix $[k]$ and using the applied nodal force matrix $[F]$, the nodal point displacement matrix $[q]$ is calculated from Eq. (11.18). The strains in the element $[\varepsilon]$ are obtained from Eq. (11.27) where matrix $[B]$ is calculated from Eq. (11.43). The corresponding uniform stress $[\sigma]$ in the element is obtained from Eq. (11.8) where $[D]$ is obtained from Eq. (11.44).

Example 11.2

The triangular plate, Figure 11.5a, is stiffened at the edges as shown. Find the stress in the various components. Let $E = 30\,000$ ksi and $\mu = 0.3$.

Solution

The various nodal points are numbered as shown in Figure 11.5b. The stiffness matrices, K, for members A, B, and C are obtained from Eq. (11.37) as

Member A with $\alpha = 0°$

With nodal points q_1, q_2, q_3, and q_4,

$$K_A = \frac{0.5 \times 3000 \times 10^4}{30} \begin{bmatrix} 1.00 & 0.00 & -1.00 & 0.00 \\ 0.00 & 0.00 & 0.00 & 0.00 \\ -1.00 & 0.00 & 1.00 & 0.00 \\ 0.00 & 0.00 & 0.00 & 0.00 \end{bmatrix}$$

or, since q_1, q_2, and q_4 are zero, we eliminate rows and columns 1, 2, and 4, and we get

$$K_A = 10^4[50].$$

Member B with $\alpha = 135°$

With nodal points q_3, q_4, q_5, and q_6,

$$K_B = \frac{1 \times 3000 \times 10^4}{28.28} \begin{bmatrix} 0.50 & -0.50 & -0.50 & 0.50 \\ -0.50 & 0.50 & 0.50 & -0.50 \\ -0.50 & 0.50 & 0.50 & -0.50 \\ 0.50 & -0.50 & -0.50 & 0.50 \end{bmatrix}$$

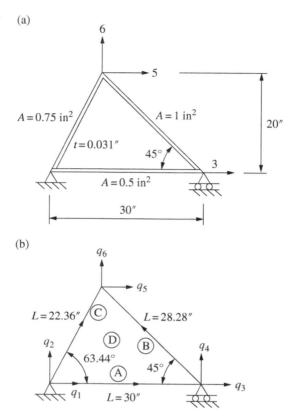

Figure 11.5 Stiffened triangular plate: (a) triangular plate with edge stiffeners and (b) nodal point designations

or, since q_4 is zero,

$$
K_B = 10^4 \begin{bmatrix} 53.04 & -53.04 & 53.04 \\ -53.04 & 53.04 & -53.04 \\ 53.04 & -53.04 & 53.04 \end{bmatrix}.
$$

Member C with $\alpha = 63.44°$

With nodal points q_1, q_2, q_5, and q_6

$$
K_C = \frac{0.75 \times 3000 \times 10^4}{22.36} \begin{bmatrix} 0.20 & 0.40 & -0.20 & -0.40 \\ 0.40 & 0.80 & -0.40 & -0.80 \\ -0.20 & -0.40 & 0.20 & 0.40 \\ -0.40 & -0.80 & 0.40 & 0.80 \end{bmatrix}
$$

or, since q_1 and q_2 are zero,

$$K_C = 10^4 \begin{bmatrix} 20.125 & 40.25 \\ 40.25 & 80.50 \end{bmatrix}.$$

Member D
With q_1, q_2, and q_4 equal to zero
From Eq. (11.45)

$$K_D = \frac{Et}{4\Delta} \begin{bmatrix} K_{33} & K_{35} & K_{36} \\ & K_{55} & K_{56} \\ & & K_{66} \end{bmatrix}$$

$$K_D = 10^4 \begin{bmatrix} 74.11 & -17.90 & 30.69 \\ -17.90 & 53.71 & 0.00 \\ 30.69 & 0.00 & 153.31 \end{bmatrix}.$$

From Table 11.2, the total matrix is

$$K = 10^4 \begin{bmatrix} 177.15 & -70.94 & 83.73 \\ -70.94 & 126.88 & -12.79 \\ 83.73 & -12.79 & 286.85 \end{bmatrix}$$

and the force matrix is

$$F = \begin{bmatrix} 3.0 \\ 5.0 \\ 7.0 \end{bmatrix} \text{ kips.}$$

From Eq. (11.32),

$$Kq = F$$

Table 11.2 Total stiffness matrix (multiplied by 10^4)

F/q	3	5	6
3	50.00		
	53.04	−53.04	53.04
	74.11	−17.90	30.69
5	−53.04	53.04	−53.04
		20.13	40.25
	−17.90	53.71	0.00
6	53.04	−53.04	53.04
		40.25	80.50
	30.69	0.0	153.31

or

$$\begin{bmatrix} q_3 \\ q_5 \\ q_6 \end{bmatrix} = \begin{bmatrix} 3.235 \\ 5.927 \\ 1.760 \end{bmatrix} \times 10^{-6} \text{inch.}$$

The stresses in members A, B, and C are obtained from Eq. (11.38) as

$$\sigma_A = 3.24 \text{ksi}, \quad \sigma_B = -0.7 \text{ksi}, \quad \sigma_C = 5.67 \text{ksi.}$$

The stress in plate D is obtained from Eq. (11.8) as

$$\begin{bmatrix} \sigma_x \\ \sigma_y \\ \tau_{xy} \end{bmatrix} = \begin{bmatrix} 8.85 \\ 7.94 \\ 5.59 \end{bmatrix} \text{ksi.}$$

11.4 Axisymmetric Triangular Linear Elements

Many plate and shell configurations, Figure 11.1, are modeled as axisymmetric triangular elements. Axisymmetric triangular elements, Figure 11.6, have the same size N matrix as that defined by Eq. (11.41) for plane elements. The strain–stress relationship given by Eq. (11.42) for plane elements must be modified for axisymmetric elements to include the hoop strain ε_θ. Thus, Eq. (11.42) becomes

$$\begin{bmatrix} \varepsilon_r \\ \varepsilon_z \\ \varepsilon_\theta \\ \gamma_{rz} \end{bmatrix} = \begin{bmatrix} \dfrac{\partial}{\partial r} & 0 \\ 0 & \dfrac{\partial}{\partial r} \\ \dfrac{1}{r} & 0 \\ \dfrac{\partial}{\partial z} & \dfrac{\partial}{\partial r} \end{bmatrix} \begin{bmatrix} u \\ v \end{bmatrix} \tag{11.46}$$

and the $[B]$ matrix becomes

$$[B] = \frac{1}{2\Delta} \begin{bmatrix} b_i & 0 & b_j & 0 & b_k & 0 \\ 0 & c_i & 0 & c_j & 0 & c_k \\ \dfrac{2\Delta N_i}{r} & 0 & \dfrac{2\Delta N_j}{r} & 0 & \dfrac{2\Delta N_k}{r} & 0 \\ c_i & b_i & c_j & b_j & c_k & b_k \end{bmatrix} \tag{11.47}$$

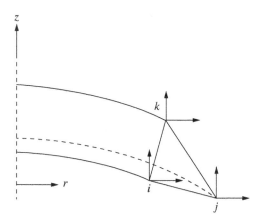

Figure 11.6 Triangular element

where

$$
\begin{bmatrix} \varepsilon_r \\ \varepsilon_z \\ \varepsilon_\theta \\ \gamma_{rz} \end{bmatrix} = [B] \begin{bmatrix} q_{ir} \\ q_{iz} \\ q_{jr} \\ q_{jz} \\ q_{kr} \\ q_{kz} \end{bmatrix}.
\tag{11.48}
$$

The stress–strain relationship is obtained from

$$
[D] = \frac{E(1-\mu)}{(1+\mu)(1-2\mu)} \begin{bmatrix} 1 & \dfrac{\mu}{1-\mu} & \dfrac{\mu}{1-\mu} & 0 \\[2mm] \dfrac{\mu}{1-\mu} & 1 & \dfrac{\mu}{1-\mu} & 0 \\[2mm] \dfrac{\mu}{1-\mu} & \dfrac{\mu}{1-\mu} & 1 & 0 \\[2mm] 0 & 0 & 0 & \dfrac{1-2\mu}{2(1-\mu)} \end{bmatrix}.
\tag{11.49}
$$

The stiffness matrix is determined from Eq. (11.16) as

$$
[K_e] = \int_v [B_e]^T [D_e][B_e]\, dV.
$$

The evaluation of the integral $(B^T DB)\, dV$ in axisymmetric problems is complicated by the fact that the matrix $[B]$ contains the variable $1/r$. A common procedure for integrating this quantity (Zienkiewicz 1977) is to use the radius $\bar{r}$ at the centroid of the element. Also, we can substitute for the quantity dV the value $(2\pi\bar{r}A)$ where A is the area of the element. Hence, the stiffness matrix $[K]$ becomes

$$[K] = [B]^T[D][B]2\bar{r}A. \tag{11.50}$$

Equations can also be derived for linear rectangular elements as shown in the top sketch in Figure 11.7. The equations for the rectangular elements are slightly more complicated than those for triangular elements (Rocky et al. 1975) due to the additional fourth nodal point. In both cases the strain, and thus stress, is constant throughout the element. This is a disadvantage in areas where a considerable strain gradient exists and a large number of linear elements are required to accurately determine this strain.

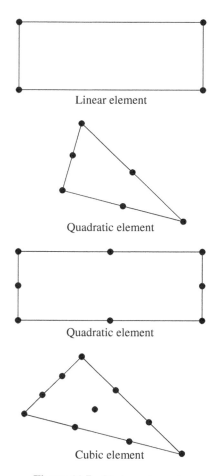

Linear element

Quadratic element

Quadratic element

Cubic element

Figure 11.7   Various elements

11.5 Higher-Order Elements

The higher-order quadratic and cubic elements, Figure 11.7, have a distinct advantage over linear elements due to the additional nodal points along the sides and middle. These additional points enable the formulation of equations that give variable stress throughout the element. With variable stress the number of elements can be reduced substantially compared with linear elements. The drawback of higher-order elements is that it takes more computer memory and storage capacity to manipulate the required equations. Accordingly, the use of these elements became popular only in the last fifteen years when computer speed and storage capacity was improved.

The shape function [N] needed to define higher-order elements is more complicated than that of linear elements, and its derivation requires more sophisticated methods using natural coordinate systems (Weaver and Johnston 1984). Also the stiffness matrix, which is a function of [N], requires numerical integration that is more cumbersome than that for linear elements. The accuracy of the results depends, in part, on the method used for the numerical integration. Derivation of the shape and stiffness matrices for higher-order elements is beyond the scope of this book. However the example shown in the following text illustrates the difference in results between a linear and a quadratic triangular element.

Finite element formulation of a plate element and finite element formulation of a shell element have also been derived (Gallagher 1975) and are based on various polynomial approximations. The accuracy of these formulations depends on the particular plate or shell theory being used.

Finite element formulation of three-dimensional brick elements is also available in the literature. The equations become fairly complex for elements higher than quadratic.

Example 11.3
Calculate the stress in the triangular element shown in Figure 11.8 using

1. Linear element theory
2. Quadratic element theory

Let $E = 30,000$ ksi, $\mu = 0.3$, $t = 1.0$ inch. Use plane stress condition. Assume the element to be laterally supported.

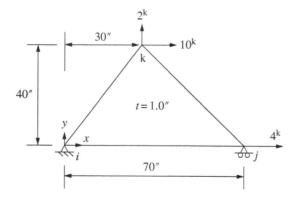

Figure 11.8   Triangular element

Solution

a. Linear element

Since nodal point i is fixed in the x- and y-directions and nodal point j is fixed in the y-direction, the stiffness matrix $[k]$ in Eq. (11.45) reduces to a 3×3 matrix with values

$$K = 10^7 \begin{bmatrix} 1.127 & -0.4327 & 4.945 \\ -0.4327 & 1.010 & 0.0 \\ 4.945 & 0.0 & 2.885 \end{bmatrix}.$$

The nodal force matrix is

$$[F] \begin{bmatrix} 4000 \\ 10,000 \\ 2000 \end{bmatrix} \text{lbs.}$$

The nodal point displacements $[q]$ are obtained from Eq. (11.18), the strain from Eq. (11.27), and the stress from Eq. (11.8)

$$[q] = 10^{-5} \begin{bmatrix} 92.67 \\ 138.8 \\ -8.952 \end{bmatrix} \text{inch}$$

$$[\varepsilon] = 10^{-5} \begin{bmatrix} 2.648 \\ -0.448 \\ 4.952 \end{bmatrix} \text{inch/inch}$$

$$[\sigma] = \begin{bmatrix} 829 \\ 114 \\ 574 \end{bmatrix} \text{psi.}$$

Notice that in a linear element, the stress in the x-direction of 829 psi, the stress in the y-direction of 114 psi, and the shear stress of 574 psi are all assumed uniform throughout the element.

b. Quadratic element

The calculations for obtaining the stress in a quadratic element generally follow the same sequence as that shown previously for the linear element. However, the actual numerical calculations are lengthier than the linear case by about a factor of ten for this case. The resultant stress distribution in the x- and y-directions is shown in Figure 11.9.

The stress pattern obtained from the quadratic element is the same as that obtained from statics for this problem. The results obtained from the linear element are off by a large margin and the element must be subdivided into smaller elements in order to obtain a more reasonable answer.

(a)

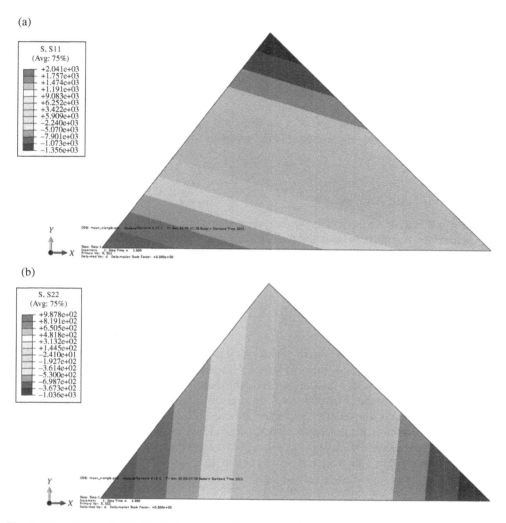

(b)

Figure 11.9   Stress distribution in the x- and y-directions for the element shown in Figure 11.8. (a) Stress in the x-direction and (b) stress in the y-direction. Source: Courtesy of Dr. Chithranjan Nadarajah.

11.6   Nonlinear Analysis

Nonlinear analysis is required to solve certain problems encountered in boiler, pressure vessel, and nuclear components. These problems include plastic analysis, buckling, and creep evaluation. In all of these cases, the solution consists of incrementing the load and adjusting various parameters to obtain a solution. Progress is still being made in available commercial programs to improve the results of such analyses.

11.6.1   Plastic Analysis

In its simplest form, elastic perfectly plastic analysis consists of incrementing the applied load and revising the stiffness matrix as the stress in various elements within the structure reach the yield value. The revision in the stiffness matrix consists of reducing the modulus of elasticity as each element reaches the specified yield value. The load is incremented until the structure collapses at which point the plastic limit is reached.

The problem becomes more complicated for strain-hardening material. In such cases the material property needs to be introduced in the solution, and the yield stress becomes variable as the strain increases.

11.6.2   Buckling Analysis

Buckling problems using finite element analysis are fairly complicated due to the fact that large displacement theory must be used and the stiffness matrix has to be modified with each increment of load in order to consider the new displaced position of the structure. The implementation of these two parameters in the solution is not easy, and at the present time, different commercial programs give slightly different answers.

11.6.3   Creep Analysis

The finite element formulation for creep analysis must consider numerous complicated variables. The first is the variation of material physical property with temperature. This is important since different elements within the structure have different temperatures. The second variable is the stress relaxation property of each element as a function of stress, time, and temperature. Obtaining such data is very costly, and at the present time, there are data for only a handful of materials. In addition, the acceptance criteria needed in deciding when a finite element analysis is completed include numerous factors such as creep limits, fatigue limits, and minimum relaxation limits. All of which are not easy to obtain. Inserting all of this data and acceptance criteria in a computer program is no easy feat.

Appendix A

Fourier Series

A.1 General Equations

A periodic function (Wylie 1972) can be represented by a series that is expressed as

$$f(x) = 0.5A_o + A_1 \cos x + A_2 \cos 2x + \cdots + A_m \cos mx + B_1 \sin x$$
$$+ B_2 \sin 2x + \cdots + B_m \sin mx$$

or

$$f(x) = 0.5A_o + \sum_{m=1}^{\infty} A_m \cos mx + \sum_{m=1}^{\infty} B_m \sin mx. \tag{A.1}$$

The series given in Eq. (A.1) is known as a Fourier series and is used to express periodic functions such as those shown in Figure A.1. The coefficients A and B in Eq. (A.1) are evaluated over a 2π period starting at a given point d. The value of A_o can be obtained by integrating Eq. (A.1) from $x = d$ to $x = d + 2\pi$.

Thus,

$$\int_d^{d+2\pi} f(x)dx = 0.5A_o \int_d^{d+2\pi} dx + A_1 \int_d^{d+2\pi} \cos x \, dx + \cdots$$

$$+ A_m \int_d^{d+2\pi} \cos mx \, dx + B_1 \int_d^{d+2\pi} \sin x \, dx + \cdots$$

$$+ B_m \int_d^{d+2\pi} \sin mx \, dx.$$

Stress in ASME Pressure Vessels, Boilers, and Nuclear Components, First Edition. Maan H. Jawad.
© 2018, The American Society of Mechanical Engineers (ASME), 2 Park Avenue,
New York, NY, 10016, USA (www.asme.org). Published 2018 by John Wiley & Sons, Inc.

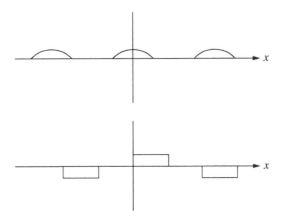

Figure A.1 Periodic functions

The first term in the right-hand side of the equation gives πA_{o}. All other terms on the right-hand side are zero because of the relationships

$$\int_d^{d+2\pi} \cos mx\ dx = 0 \quad m \neq 0$$

$$\int_d^{d+2\pi} \sin mx\ dx = 0.$$

Hence,

$$A_{\mathrm{o}} = \frac{1}{\pi}\int_d^{d+2\pi} f(x)dx. \tag{A.2}$$

The A_m term in Eq. (A.1) can be obtained by multiplying both sides of the equation by $\cos mx$.

$$\int_d^{d+2\pi} f(x)\cos mx\ dx = \frac{1}{2}A_{\mathrm{o}}\int_d^{d+2\pi} \cos mx\ dx + A_1\int_d^{d+2\pi} \cos x\ \cos mx\ dx + \cdots$$

$$+ A_m\int_d^{d+2\pi} \cos mx\ \cos mx\ dx + B_1\int_d^{d+2\pi} \sin x\ \cos mx\ dx + \cdots \tag{A.3}$$

$$+ B_m\int_d^{d+2\pi} \sin mx\ \cos mx\ dx.$$

Since

$$\int_d^{d+2\pi} \cos mx \cos nx\ dx = 0 \quad m \neq n$$

$$\int_d^{d+2\pi} \cos^2 mx\ dx = \pi \quad m \neq 0$$

$$\int_d^{d+2\pi} \cos mx \sin nx\ dx = 0$$

Equation (A.3) becomes

$$\int_d^{d+2\pi} f(x)\cos mx \; dx = A_m \pi$$

or

$$A_m = \frac{1}{\pi}\int_d^{d+2\pi} f(x)\cos mx \; dx. \tag{A.4}$$

Similarly the values of B_m can be found by multiplying both sides of Eq. (A.1) by sin mx. Using the expressions

$$\int_d^{d+2\pi} \sin mx \sin nx \; dx = 0 \;\; m \neq n$$

and

$$\int_d^{d+2\pi} \sin^2 mx \; dx = \pi,$$

the equation becomes

$$B_m = \frac{1}{\pi}\int_d^{d+2\pi} f(x)\sin \; mx \; dx. \tag{A.5}$$

Accordingly, we can state that for a given periodic function $f(x)$, a Fourier expansion can be written as shown in Eq. (A.1) with the various constants obtained from Eqs. (A.2), (A.4), and (A.5).

Example A.1
Express the function shown in Figure A.2 in a Fourier series.

Solution

$$f(x) = 0 \;\; -\pi < x < 0$$
$$f(x) = p_0 \;\; 0 < x < \pi$$
$$d = -\pi$$

From Eq. (A.2),

$$A_0 = \frac{1}{\pi}\int_{-\pi}^0 (0)dx + \frac{1}{\pi}\int_0^\pi p_0 dx$$

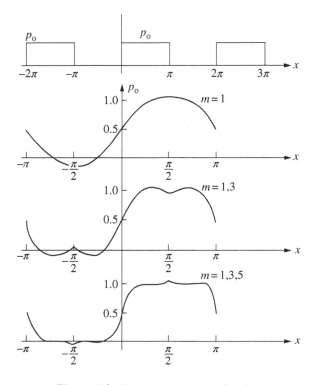

Figure A.2   Periodic rectangular function

or

$$A_\mathrm{o} = p_\mathrm{o}.$$

From Eq. (A.4),

$$A_m = \frac{1}{\pi} \int_0^\pi p_\mathrm{o} \cos mx \ dx$$

$$A_m = 0.$$

From Eq. (A.5),

$$B_m = \frac{1}{\pi} \int_0^\pi p_o \sin mx \ dx$$

$$= \frac{p_\mathrm{o}}{m\pi} (-\cos mx)\Big|_0^\pi$$

$$= \frac{-p_\mathrm{o}}{m\pi} (\cos m\pi - 1)$$

$$B_m = \frac{2p_\mathrm{o}}{m\pi} \quad \text{when} \quad m \text{ is odd}$$

$$= 0 \quad \text{when} \quad m \text{ is even.}$$

Therefore, the expansion of the function shown in Figure A.2 is expressed as

$$f(x) = 0.5p_0 + \frac{2p_0}{\pi} \sum_{m=1,3,\ldots}^{\infty} \frac{1}{m} \sin mx.$$

A plot of this equation with $m = 1, 3, 5$ is shown in Figure A.2.

A.2 Interval Change

In applying the Fourier series to plate and shell problems, it is more convenient to specify intervals other than 2π. Defining the new interval as $2p$, Eqs. (A.2), (A.4), and (A.5) can be written as

$$A_0 = \frac{1}{p} \int_d^{d+2p} f(x) dx \tag{A.6}$$

$$A_m = \frac{1}{p} \int_d^{d+2p} f(x) \cos \frac{m\pi x}{p} dx \tag{A.7}$$

$$B_m = \frac{1}{p} \int_d^{d+2p} f(x) \sin \frac{m\pi x}{p} dx \tag{A.8}$$

where $2p = $ period of function, and the series can be written as

$$f(x) = \frac{1}{2}A_0 + \sum_{m=1}^{\infty} A_m \cos \frac{m\pi x}{p} + \sum_{m=1}^{\infty} B_m \sin \frac{m\pi x}{p}. \tag{A.9}$$

Example A.2
Find the Fourier expansion of the function $f(x) = \cos x$ as shown in Figure A.3.

Solution
The period $2p$ is equal to π. Thus, $p = \pi/2$ and $d = -\pi/2$.

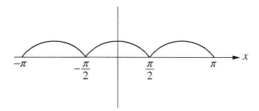

Figure A.3   Periodic cosine function

$$A_o = \frac{1}{\pi/2} \int_{-\pi/2}^{\pi/2} \cos x \, dx = \frac{4}{\pi}$$

$$A_m = \frac{2}{\pi} \int_{-\pi/2}^{\pi/2} \cos x \cos \frac{m\pi x}{\pi/2} dx = \frac{4}{\pi} \sum_{m=1}^{\infty} \frac{(-1)^{m+1}}{(4m^2-1)}$$

$$B_m = 0.$$

and from Eq. (A.9),

$$f(x) = \frac{2}{\pi} + \frac{4}{\pi} \sum_{m=1}^{\infty} \frac{(-1)^{m+1}}{4m^2-1} \cos 2mx.$$

A.3   Half-Range Expansions

If a function is symmetric with respect to the axis of reference as shown in Figure A.4, then the coefficient integral can be simplified by integrating over one-half the period. This integration can be performed as an even or odd function.

Hence, for an even periodic function,

$$
\left.\begin{aligned}
A_o &= \frac{2}{p} \int_0^p f(x) dx \\
A_m &= \frac{2}{p} \int_0^p f(x) \cos \frac{m\pi x}{p} dx \\
B_m &= 0.
\end{aligned}\right]
\tag{A.10}
$$

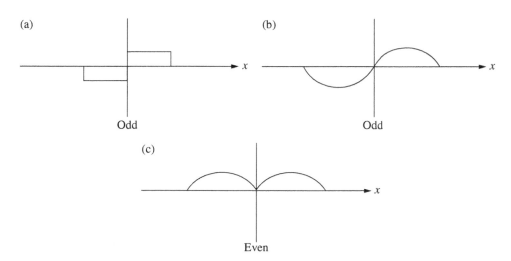

Figure A.4 Periodic symmetric function. (a) Odd step function, (b) odd continuous function, and (c) even step function

For an odd periodic function,

$$A_o = A_m = 0$$

$$\left. B_m = \frac{2}{p}\int_0^p f(x)\sin\frac{m\pi x}{p}dx \right]$$

(A.11)

It should be noted that the even and odd functions defined by Eqs. (A.10) and (A.11) and Figure A.4 do not refer necessarily to the shape of the function but rather to the reference line from which they are defined. This can best be illustrated by Example A.3.

Example A.3
Figure A.5 shows a plot of the function $y = x - x^2$. Obtain and plot the Fourier series expansion of this function (a) from $y = -1$ to $y = 1$; (b) as an even series from $y = 0$ to $y = 1$; and (c) as an odd series from $y = 0$ to $y = 1$.

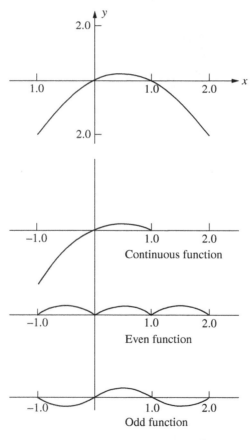

Figure A.5   Function $y = x - x^2$

Solution

a.
$$d = -1, \quad 2p = 2 \quad \text{or} \quad p = 1$$

$$A_0 = \int_{-1}^{1} (x - x^2) dx = -\frac{2}{3}$$

$$A_m = \int_{-1}^{1} (x - x^2) \cos \frac{m\pi x}{1} dx = -\frac{4 \cos m\pi}{m^2 \pi^2}$$

$$B_m = \int_{-1}^{1} (x - x^2) \sin \frac{m\pi x}{1} dx = -\frac{2 \cos m\pi}{m\pi}$$

$$f(x) = -\frac{1}{3} - \sum_{m=1}^{\infty} \frac{4 \cos m\pi}{m^2 \pi^2} \cos m\pi x - \sum_{m=1}^{\infty} \frac{2 \cos m\pi}{m\pi} \sin m\pi x$$

$$= -\frac{1}{3} - \frac{4}{\pi^2} \sum_{m=1}^{\infty} \frac{(-1)^m}{m^2} \cos m\pi x - \frac{2}{\pi} \sum_{m=1}^{\infty} \frac{(-1)^m}{m} \sin m\pi x$$

b.
$$A_0 = 2 \int_{0}^{1} (x - x^2) dx = \frac{1}{3}$$

$$A_m = 2 \int_{0}^{1} (x - x^2) \cos \frac{m\pi x}{1} dx = -\frac{2(1 + \cos m\pi)}{m^2 \pi^2}$$

$$B_m = 0$$

$$f(x) = \frac{1}{6} - \sum_{m=1}^{\infty} \frac{2(1 + \cos m\pi)}{m^2 \pi^2} \cos m\pi x$$

c.
$$A_0 = A_m = 0$$

$$B_m = 2 \int_{0}^{1} (x - x^2) \sin \frac{m\pi x}{1} dx = \frac{4(1 - \cos m\pi)}{m^3 \pi^3}$$

$$f(x) = \sum_{m=1}^{\infty} \frac{4(1 - \cos m\pi)}{m^3 \pi^3} \sin m\pi x$$

A.4   Double Fourier Series

In solving rectangular plate problems of length a and width b, it is customary to express the applied loads in terms of a single or double series. The double Fourier series is normally expressed as an odd periodic function with a half-range period given between 0 and a for one side of the plate and 0 and b for the other side. Thus,

$$f(x,y) = \sum_{m=1}^{\infty} \sum_{n=1}^{\infty} B_{mn} \sin\frac{m\pi x}{a} \sin\frac{n\pi y}{b} \tag{A.12}$$

where

$$B_{mn} = \frac{4}{ab} \int_0^b \int_0^a f(x,y) \sin\frac{m\pi x}{a} \sin\frac{n\pi y}{b} dxdy. \tag{A.13}$$

Example A.4
The rectangular plate shown in Figure A.6 is subjected to a uniform pressure p_o. Determine the Fourier expansion for the pressure.

Solution
From Eq. (A.13),

$$B_{mn} = \frac{4p_o}{ab} \int_0^b \int_0^a \sin\frac{m\pi x}{a} \sin\frac{n\pi y}{b} dxdy$$

$$= \frac{16p_o}{\pi^2 mn} \quad m,n \text{ are odd functions}$$

$$f(x,y) = \frac{16p_o}{\pi^2} \sum_{m=1,3,\ldots}^{\infty} \sum_{n=1,3,\ldots}^{\infty} \frac{1}{mn} \sin\frac{m\pi x}{a} \sin\frac{n\pi y}{b}$$

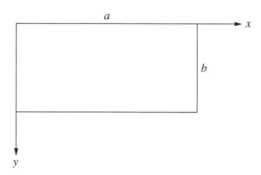

Figure A.6   Rectangular plate

Appendix B

Bessel Functions

B.1 General Equations

In many plate and shell applications involving circular symmetry, the resulting differential equations are solved by means of a power series known as Bessel functions. Some of these functions are discussed in this appendix.

The differential equation

$$\frac{d^2y}{dx^2} + \frac{1}{x}\frac{dy}{dx} + y = 0 \tag{B.1}$$

is referred to as Bessel's equation of zero order. Its solution (Bowman 1977) is given by the following power series:

$$y = C_1 J_0(x) + C_2 Y_0(x) \tag{B.2}$$

where C_1 and $C_2 =$ constants obtained from boundary conditions and $J_0(x) =$ Bessel function of the first kind of zero order.

$$J_0(x) = 1 - \frac{x^2}{2^2} + \frac{x^4}{2^2 \cdot 4^2} - \frac{x^6}{2^2 \cdot 4^2 \cdot 6^2} + \cdots$$

$$= \sum_{m=0}^{\infty} \frac{(-1)^m}{(m!)^2} \left(\frac{x}{2}\right)^{2m}$$

$Y_0(x) =$ Bessel function of the second kind of zero order.

Stress in ASME Pressure Vessels, Boilers, and Nuclear Components, First Edition. Maan H. Jawad.
© 2018, The American Society of Mechanical Engineers (ASME), 2 Park Avenue,
New York, NY, 10016, USA (www.asme.org). Published 2018 by John Wiley & Sons, Inc.

$$Y_o(x) = J_o(x) \int \frac{dx}{x J_o^2(x)}$$

$$= J_o(x)\ln x + \frac{x^2}{2^2} - \frac{x^4}{2^2 \cdot 4^2}\left(1 + \frac{1}{2}\right) + \frac{x^6}{2^2 \cdot 4^2 \cdot 6^2}\left(1 + \frac{1}{2} + \frac{1}{3}\right) - \cdots$$

Equation (B.1) is usually encountered in a more general form as

$$x^2 \frac{d^2 y}{dx^2} + x \frac{dy}{dx} + (x^2 - k^2)y = 0. \qquad (B.3)$$

The solution of this equation (Hildebrand 1964) is

$$y = C_1 J_k(x) + C_2 J_{-k}(x)$$

when k is not zero or a positive integer or

$$y = C_1 J_k(x) + C_2 Y_k(x)$$

when k is zero or a positive integer and where

$J_k(x) =$ Bessel function of the first kind of order k

$$= \sum_{m=0}^{\infty} \frac{(-1)^m}{(m!)(m+k)!}\left(\frac{x}{2}\right)^{2m+k}$$

$J_{-k}(x) =$ Bessel function of the first kind of order k

$$= \sum_{m=0}^{\infty} \frac{(-1)^m}{(m!)(m-k)!}\left(\frac{x}{2}\right)^{2m-k}$$

$Y_k(x) =$ Bessel function of the second kind of order k

$$= \frac{2}{\pi}\left[\left(\ln\left(\frac{x}{2}\right) + \gamma\right)J_k(x) - \frac{1}{2}\sum_{m=0}^{k-1}\frac{(k-m-1)!}{m!}\left(\frac{x}{2}\right)^{2m-k}\right.$$

$$\left. + \frac{1}{2}\sum_{m=0}^{\infty}(-1)^{m+1}[h(m) + h(m+k)]\frac{(x/2)^{2m+k}}{m!(m+k)!}\right]$$

$$h(m) = \sum_{r=1}^{m}\frac{1}{r} \quad m > 1.$$

$$\gamma = 0.5772$$

A plot of $J_o(x)$, $J_1(x)$, $Y_o(x)$, and $Y_1(x)$ is shown in Figure B.1. Also, Table B.1 gives some values of $J(x)$ and $Y(x)$.

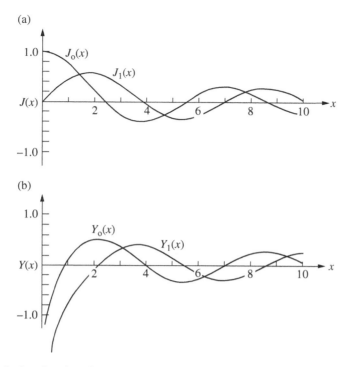

Figure B.1 A plot of various Bessel functions: (a) J_o and J_1 functions and (b) Y_o and Y_1 functions

Table B.1 Values of J_o, J_1, Y_o, and Y_1

(x)	$J_o(x)$	$J_1(x)$	$Y_o(x)$	$Y_1(x)$
0.0	1.0000	0.0000	$-\infty$	$-\infty$
0.5	0.9385	0.2423	−0.4445	−1.4715
1.0	0.7652	0.4401	0.0883	−0.7812
1.5	0.5118	0.5579	0.3825	−0.4123
2.0	0.2239	0.5767	0.5104	−0.1070
2.5	−0.0484	0.4971	0.4981	0.1459
3.0	−0.2601	0.3391	0.3769	0.3247
3.5	−0.3801	0.1374	0.1890	0.4102
4.0	−0.3972	−0.0660	−0.0169	0.3979
4.5	−0.3205	−0.2311	−0.1947	0.3010
5.0	−0.1776	−0.3276	−0.3085	0.1479
5.5	−0.0068	−0.3414	−0.3395	−0.0238
6.0	0.1507	−0.2767	−0.2882	−0.1750
6.5	0.2601	−0.1538	−0.1732	−0.2741
7.0	0.3001	−0.0047	−0.0260	−0.3027
7.5	0.2663	0.1353	0.1173	−0.2591
8.0	0.1717	0.2346	0.2235	−0.1581
8.5	0.0419	0.2731	0.2702	−0.0262
9.0	−0.0903	0.2453	0.2499	0.1043
9.5	−0.1939	0.1613	0.1712	0.2032
10.0	−0.2459	0.0435	0.0557	0.2490

A different form of Eq. (B.3) that is encountered often in plate and shell theory is

$$x^2\frac{d^2y}{dx^2} + x\frac{dy}{dx} - (x^2 + k^2)y = 0.$$ (B.4)

The solution of this equation (Dwight 1972) is

$$y = C_1 I_k(x) + C_2 I_{-k}(x)$$

when k is not zero or a positive integer or

$$y = C_1 I_k(x) + C_2 K_k(x)$$

when k is zero or a positive integer and where

$I_k(x) =$ modified Bessel function of the first kind of order k

$$= \sum_{m=1}^{\infty} \frac{(x/2)^{2m+k}}{m!(m+k)!}.$$

$K_k(x) =$ modified Bessel function of the second kind of order k

$$= (-1)^{k+1}\left[\ln\left(\frac{x}{2}\right) + \gamma\right]I_k(x)$$

$$+ \frac{1}{2}\sum_{m=0}^{k-1}\frac{(-1)^m(k-m-1)!}{m!}\left(\frac{x}{2}\right)^{2m-k}$$

$$+ \frac{1}{2}\sum_{m=0}^{\infty}\frac{(-1)^k(x/2)^{2m+k}}{m!(m+k)!}\left[\left(1 + \frac{1}{2} + \cdots + \frac{1}{m}\right) + \left(1 + \frac{1}{2} + \cdots + \frac{1}{m+k}\right)\right].$$

A plot of $I_0(x)$, $I_1(x)$, $K_0(x)$, and $K_1(x)$ is shown in Figure B.2.

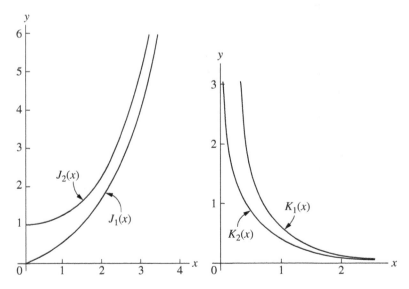

Figure B.2   A plot of modified Bessel functions (Wylie 1972)

Another equation that is often encountered in plate and shell theory is given by

$$x^2 \frac{d^2y}{dx^2} + x\frac{dy}{dx} - (ix^2 + k^2)y = 0. \tag{B.5}$$

The solution of this equation (Hetenyi 1964) for the important case of $k = 0$ is given by

$$y = C_1 Z_1(x) + C_2 Z_2(x) + C_3 Z_3(x) + C_4 Z_4(x) \tag{B.6}$$

where

$$Z_1(x) = \mathrm{ber}(x) = \sum_{m=0}^{\infty}(-1)^m \frac{(x/2)^{4m}}{[(2m)!]^2}$$

$$Z_2(x) = -\mathrm{bei}(x) = -\sum_{m=0}^{\infty}(-1)^m \frac{(x/2)^{4m+2}}{[(2m+1)!]^2}$$

$$Z_3(x) = -\frac{2}{\pi}\mathrm{kei}(x) = \frac{Z_1(x)}{2} - \frac{2}{\pi}\left[R_1 + \left(\ln\left(\frac{\gamma x}{2}\right)\right)(Z_2(x))\right]$$

$$Z_4(x) = -\frac{2}{\pi}\mathrm{ker}(x) = \frac{Z_2(x)}{2} - \frac{2}{\pi}\left[R_2 + \left(\ln\left(\frac{\gamma x}{2}\right)\right)(Z_1(x))\right]$$

$$R_1 = \left(\frac{x}{2}\right)^2 - \frac{h(3)}{(3!)^2}\left(\frac{x}{2}\right)^6 + \frac{h(5)}{(5!)^2}\left(\frac{x}{2}\right)^{10} - \cdots$$

$$R_2 = \frac{h(2)}{(2!)^2}\left(\frac{x}{2}\right)^4 - \frac{h(4)}{(4!)^2}\left(\frac{x}{2}\right)^8 + \frac{h(6)}{(6!)^2}\left(\frac{x}{2}\right)^{12} - \cdots$$

$$h(n) = 1 + \frac{1}{2} + \frac{1}{3} + \cdots + \frac{1}{n}$$

$$\gamma = 0.5772.$$

A plot of $Z_1(x)$, $Z_2(x)$, $Z_3(x)$, $Z_4(x)$, and their derivatives is shown in Figure B.3.

B.2   Some Bessel Identities

The derivatives and integrals of Bessel functions follow a certain pattern. The identities given here are needed to solve some of the problems given in this text.

$$\frac{d}{dx}[xJ_1(x)] = xJ_0(x)$$

$$\frac{d}{dx}J_0(x) = -J_1(x)$$

$$\frac{d}{dx}[x^n J_n(x)] = x^n J_{n-1}(x)$$

$$\frac{d}{dx}\left[\frac{J_n(x)}{x^n}\right] = \frac{-J_{n+1}(x)}{x^n}$$

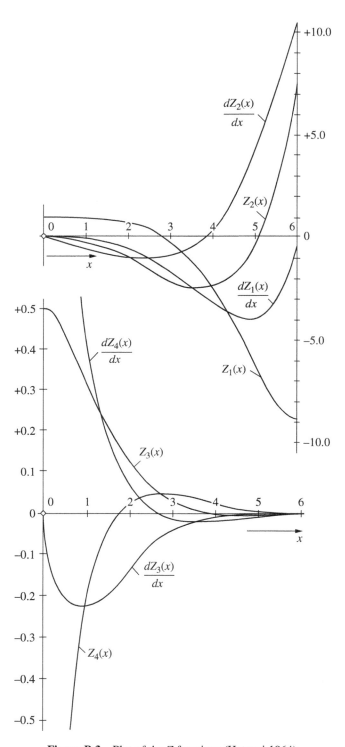

Figure B.3   Plot of the Z functions (Hetenyi 1964)

$$\frac{d^2 Z_1(x)}{dx^2} = Z_2(x) - \frac{1}{x}\frac{dZ_1(x)}{dx}$$

$$\frac{d^2 Z_2(x)}{dx^2} = -Z_1(x) - \frac{1}{x}\frac{dZ_2(x)}{dx}$$

$$\frac{d^2 Z_3(x)}{dx^2} = Z_4(x) - \frac{1}{x}\frac{dZ_3(x)}{dx}$$

$$\frac{d^2 Z_4(x)}{dx^2} = -Z_3(x) - \frac{1}{x}\frac{dZ_4(x)}{dx}$$

The last four equations are needed in the solution of circular plates on elastic foundation. In these equations the value of (kx) is needed rather than (x) in the Z functions. In this case, these equations take on the form

$$k^2 Z_1''(kx) = k^2 Z_2(x) - \frac{k}{x}Z_1'(kx)$$

$$k^2 Z_2''(kx) = -k^2 Z_1(x) - \frac{k}{x}Z_2'(kx)$$

$$k^2 Z_3''(kx) = k^2 Z_4(x) - \frac{k}{x}Z_3'(kx)$$

$$k^2 Z_4''(kx) = -k^2 Z_3(x) - \frac{k}{x}Z_4'(kx)$$

B.3   Simplified Bessel Functions

As x approaches zero, the various Bessel functions can be expressed as

$$J_k(x) = \frac{x^k}{(2^k)(k!)}$$

$$Y_k(x) = \frac{-2^k(k-1)!}{\pi}x^{-k} \quad k \neq 0$$

$$Y_o(x) = \frac{2}{\pi}\ln x$$

$$I_k(x) = \frac{x^k}{2^k k!}$$

$$K_k(x) = (2^{k-1})[(k-1)!]x^{-k} \quad k \neq 0$$

$$K_o(x) = -\ln x$$

$$Z_1(x) = 1.0 \quad Z_2(x) = -\frac{x^2}{4}$$

$$Z_3(x) = \frac{1}{2} \quad Z_4(x) = \frac{2}{\pi}\ln\frac{\gamma x}{2}$$

$$\frac{dZ_1(x)}{dx} = -\frac{x^3}{16} \quad \frac{dZ_3(x)}{dx} = \frac{x}{\pi} \ln \frac{\gamma x}{2}$$

$$\frac{dZ_2(x)}{dx} = -\frac{x}{2} \quad \frac{dZ_4(x)}{dx} = \frac{2}{\pi x}$$

As x approaches infinity, the various Bessel functions can be expressed as

$$J_k(x) = \sqrt{\frac{2}{\pi x}} \cos(x - \xi_k) \quad \xi_k = (2k+1)\frac{\pi}{4}$$

$$Y_k(x) = \sqrt{\frac{2}{\pi x}} \sin(x - \xi_k)$$

$$I_k(x) = \frac{e^x}{\sqrt{2\pi x}}$$

$$K_k(x) = \frac{e^{-x}}{\sqrt{2x/\pi}}$$

$$Z_1(x) = \eta \cos\sigma \quad Z_2(x) = -\eta \sin\sigma$$

$$Z_3(x) = \beta \sin\tau \quad Z_4(x) = -\beta \cos\tau$$

$$\eta = \frac{1}{\sqrt{2\pi x}} e^{x/\sqrt{2}} \quad \beta = \sqrt{\frac{2}{\pi x}} e^{-x/\sqrt{2}}$$

$$\sigma = \frac{x}{\sqrt{2}} - \frac{\pi}{8} \quad \tau = \frac{x}{\sqrt{2}} + \frac{\pi}{8}$$

$$\frac{dZ_1(x)}{dx} = \frac{\eta}{\sqrt{2}}(\cos\sigma - \sin\sigma)$$

$$\frac{dZ_2(x)}{dx} = \frac{-\eta}{\sqrt{2}}(\cos\sigma + \sin\sigma)$$

$$\frac{dZ_3(x)}{dx} = \frac{\beta}{\sqrt{2}}(\cos\tau - \sin\tau)$$

$$\frac{dZ_4(x)}{dx} = \frac{-\beta}{\sqrt{2}}(\cos\tau + \sin\tau)$$

Appendix C

Conversion Factors

Pressure units

	1 psi	1 N/mm^2	1 bar	1 kPa	1 kgf/cm^2
psi	1.0000	145.0	14.50	0.1450	14.22
N/mm^2	0.006895	1.000	0.1000	0.0010	0.09807
bars	0.06895	10.000	1.000	0.0100	0.9807
kPa	6.895	1000.0	100.00	1.000	98.07
kgf/cm^2	0.0703	10.20	1.020	0.0102	1.000

1 N/mm^2 = 1 MPa.

Stress units

	1 ksi	1 kN/mm^2	1 MPa	1 kgf/mm^2
ksi	1.000	145.0	0.1450	1.422
kN/mm^2	0.006895	1.000	0.001	0.009807
MPa	6.895	1000.00	1.000	9.807
kgf/mm^2	0.7033	102.0	0.1020	1.000

Force units

	1 lb	1 kgf	1 N
lb	1.000	2.205	0.2248
kgf	0.454	1.000	0.1020
N	4.448	9.807	1.0000

Stress in ASME Pressure Vessels, Boilers, and Nuclear Components, First Edition. Maan H. Jawad.
© 2018, The American Society of Mechanical Engineers (ASME), 2 Park Avenue,
New York, NY, 10016, USA (www.asme.org). Published 2018 by John Wiley & Sons, Inc.

References

American Association of State Highway and Transportation Officials. 1996. Standard Specifications for Highway Bridges HB-16. Washington, DC: AASHTO.

American Concrete Institute, 1981. Concrete Shell Buckling, Publication SP-67. P. Seide, Stability of cylindrical reinforced concrete shells, SP 67–2. Detroit, Michigan.

American Institute of Steel Construction. 2013. Manual of Steel Construction-Allowable Stress Design. Chicago, IL: AISC.

American Iron and Steel Institute. 1981. Steel Penstocks and Tunnel Liners. Washington, DC: AISI.

American Petroleum Institute. 2014. Design and Construction of Large, Welded, Low-Pressure Storage Tanks—API 620. Washington, DC: API.

American Society of Mechanical Engineers. 2017a. Pressure Vessel Code, Section VIII, Division 1. New York, NY: ASME.

American Society of Mechanical Engineers. 2017b. Pressure Vessel-Alternate Rules, Section VIII, Division 2. New York, NY: ASME.

Baker, E. H., Cappelli, A. P., Kovalevsky, L., Rish, F. L., and Verette, R. M. 1968. Shell Analysis Manual—NASA 912. Washington, DC: National Aeronautics and Space Administration.

Becker, H. July 1957. Handbook of Structural Stability—Part II—Buckling of Composite Elements NACA PB 128 305. Washington, DC: National Advisory Committee for Aeronautics.

Bloom, F. and Coffin, D. 2001. Handbook of Thin Plate Buckling and Postbuckling. Boca Raton, FL: Chapman and Hall/CRC.

Boley, A. B. and Weiner, J. H. 1997. Theory of Thermal Stresses. New York, NY: Dover Publications.

Bowman, F. 1977. Introduction to Bessel Functions. New York, NY: Dover Publications.

Buchert, K. P. 1964. Stiffened Thin Shell Domes. AISC Engineering Journal, Vol. 1, pp. 78–82.

Buchert, K. P. 1966. Buckling Considerations in the Design and Construction of Doubly Curved Space Structures. International Conference on Space Structures, 1966—F8. England: University of Surry.

Burgreen, D. 1971. Elements of Thermal Stress Analysis. Jamaica, NY: C.P. Press.

Dwight, H. B. 1972. Tables of Integrals and Other Mathematical Data. New York, NY: Macmillan.

Farr, J.R. and Jawad, M.H. 2010. Guidebook for the Design of ASME Section VIII Pressure Vessels. New York, NY: ASME Press.

Faupel, J. H. and Fisher, F. E. 1981. Engineering Design. New York, NY: Wiley-Interscience.

Flugge, W. 1967. Stresses in Shells. New York, NY: Springer-Verlag.

Stress in ASME Pressure Vessels, Boilers, and Nuclear Components, First Edition. Maan H. Jawad.
© 2018, The American Society of Mechanical Engineers (ASME), 2 Park Avenue,
New York, NY, 10016, USA (www.asme.org). Published 2018 by John Wiley & Sons, Inc.

Gallagher, R. H. 1975. Finite Element Analysis. Englewood Cliffs, NJ: Prentice Hall.

Gerard, G. August 1957a. Handbook of Structural Stability—Part IV—Failure of Plates and Composite Elements. NACA N62-55784. Washington, DC: National Advisory Committee for Aeronautics.

Gerard, G. August 1957b. Handbook of Structural Stability—Part V—Compressive Strength of Flat Stiffened Panels. NACA PB 185 629. Washington, DC: National Advisory Committee for Aeronautics.

Gerard, G. 1962. Introduction to Structural Stability Theory. New York, NY: McGraw Hill.

Gerard, G. and Becker, H. July 1957a. Handbook of Structural Stability—Part I—Buckling of Flat Elements. NACA PB 185 628. Washington, DC: National Advisory Committee for Aeronautics.

Gerard, G. and Becker, H. 1957b. Handbook of Structural Stability—Part III—Buckling of Curved Plates and Shells. NACA TN 3783. Washington, DC: National Advisory Committee for Aeronautics.

Gibson, J. E. 1965. Linear Elastic Theory of Thin Shells. New York, NY: Pergamon Press.

Grandin, H. Jr. 1986. Fundamentals of the Finite Element Method. New York, NY: Macmillan.

Harrenstien, H. P. and Alsmeyer, W. C. 1959. Structural Behavior of a Plate Resembling a Constant Thickness Bridge Abutment Wingwall. Engineering Experiment Station Bulletin 182. Ames, IA: Iowa State University.

Hetenyi, M. 1964. Beams on Elastic Foundation. Ann Arbor, MI: University of Michigan Press.

Hildebrand, F. 1964. Advanced Calculus for Applications. Englewood Cliffs, NJ: Prentice Hall.

Hult, J. H. 1966. Creep in Engineering Structures. London: Blaisdell Publishing Company.

Iyengar, N. G. R. 1988. Structural Stability of Columns and Plates. New York, NY: John Wiley & Sons, Inc.

Jawad, M. H. 1980. Design of Conical Shells Under External Pressure. Journal of Pressure Vessel Technology, Vol. 102, pp. 230–238.

Jawad, M. H. 2004. Design of Plate and Shell Structures. New York, NY: ASME Press.

Jawad, M. H. and Farr, J. R. 1989. Structural Analysis and Design of Process Equipment. New York, NY: John Wiley & Sons, Inc.

Jawad, M. H. and Jetter, R. I. 2011. Design and Analysis of ASME Boiler and Pressure Vessel Components in the Creep Range. New York, NY: ASME Press.

Jones, R. M. 2009. Deformation Theory of Plasticity. Blacksburg, VA: Bull Ridge Publishing.

Koiter, W. T. 1943. The Effective Width of Flat Plates for Various Longitudinal Edge Conditions at Loads Far Beyond the Buckling Load. Report No. 5287. The Netherlands: National Luchtvaart Laboratorium.

Kraus, H. 1967. Thin Elastic Shells. New York, NY: John Wiley & Sons, Inc.

Love, A. E. H. 1944. A Treatise on the Mathematical Theory of Elasticity. New York, NY: Dover Publications.

The M. W. Kellogg Company. 1961. Design of Piping Systems. New York, NY: John Wiley & Sons, Inc.

Marguerre, K. 1937. The Apparent Width of the Plate in Compression. Technical Memo No. 833. Washington, DC: National Advisory Committee for Aeronautics.

Miller, C. D. 1999. External Pressure. WRC Bulletin 443. New York, NY: Welding Research Council.

Niordson, F. I. N. 1947. Buckling of Conical Shells Subjected to Uniform External Lateral Pressure. Transactions of the Royal Institute of Technology, No. 10. Stockholm: Royal Institute of Technology.

O'Donnell, W. J. and Langer, B. F. 1962. Design of Perforated Plates. Journal of Engineering for Industry, Vol. 84, pp. 307–319.

Perry, C. L. 1950. The Bending of Thin Elliptic Plates. Proceedings of Symposia in Applied Mathematics, Volume III, p. 131. New York, NY: McGraw Hill.

Raetz, R. V. 1957. An Experimental Investigation of the Strength of Small-Scale Conical Reducer Sections Between Cylindrical Shells Under External Hydrostatic Pressure. U.S. Department of the Navy, David Taylor Model Basin. Report No. 1187. Washington, DC: U.S. Navy.

Roark, R. J., and Young, W. C. 1975. Formulas for Stress and Strain. New York, NY: McGraw Hill.

Rocky, K. C., Evans, H. R., Griffiths, D. W., and Nethercot, D. A. 1975. The Finite Element Method. New York, NY: John Wiley & Sons, Inc.

Segerlind, L. J. 1976. Applied Finite Element Analysis. New York, NY: John Wiley & Sons, Inc.

Seide, P. 1962. A Survey of Buckling Theory and Experiment for Circular Conical Shells of Constant Thickness. NASA Publication IND-1510. Cleveland, OH: NASA.

Seide, P. 1981. Stability of Cylindrical Reinforced Concrete Shells. In Concrete Shell Buckling. ACI SP-67. Chicago, IL: American Concrete Institute.

Shenk, A. 1997. Calculus and Analytic Geometry. Palo Alto, CA: Addison-Wesley Publishing.

Sherbourne, A. N. 1961. Elastic Postbuckling Behavior of a Simply Supported Circular Plate. Journal of Mechanical Engineering Science, Vol. 3, pp. 133–141.

Sokolnikoff, I. S. 1956. Mathematical Theory of Elasticity. New York, NY: McGraw Hill.

Sturm, R. G. 1941 A Study of the Collapsing Pressure of Thin-Walled Cylinders. Engineering Experiment Station Bulletin 329. Urbana, IL: The University of Illinois.

Swanson, H. S., Chapton, H. J., Wilkinson, W. J., King, C. L., and Nelson, E. D. June 1955. Design of Wye Branches for Steel Pipe. Journal of AWWA, Vol. 47, pp. 581–630.

Szilard, R. 1974. Theory and Analysis of Plates-Classical and Numerical Methods. Englewood Cliffs, NJ: Prentice Hall.

Timoshenko, S. P. 1983. History of Strength of Materials. New York, NY: Dover Publications.

Timoshenko, S. P. and Gere, J. M. 1961. Theory of Elastic Stability. New York, NY: McGraw Hill.

Timoshenko, S. P. and Woinowsky-Krieger, S. 1959. Theory of Plates and Shells. New York, NY: McGraw Hill.

Tubular Exchanger Manufacturers Association. 2007. Standards of the Tubular Exchanger Manufacturers Association. Tarrytown, NY: TEMA.

Ugural, A. C. 1998. Stresses in Plates and Shells, New York, NY: McGraw Hill.

Von Karman, Th. and Tsien, Hsue-Shen. 1939. The Buckling of Spherical Shells by External Pressure. In Pressure Vessel and Piping Design-Collected Papers 1927–1959. New York, NY: ASME.

Wang, C. K. 1986. Structural Analysis on Microcomputers. New York, NY: Macmillan.

Weaver, W. Jr. and Johnston, P. R. 1984. Finite Elements for Structural Analysis. Englewood Cliffs, NJ: Prentice Hall.

Wylie, C. R. Jr. 1972. Advanced Engineering Mathematics. New York, NY: John Wiley & Sons, Inc.

Young, W. C., Budynas, R. G., and Sadegh, A. M. 2012. Roark's Formulas for Stress and Strain. New York, NY: McGraw Hill.

Zick, L. P. and St. Germain, A. R. May 1963. Circumferential Stresses in Pressure Vessel Shells of Revolution. Journal of Engineering for Industry, Vol. 85, pp. 201–216.

Zienkiewicz, O. C. 1977. The Finite Element Method. New York, NY: McGraw Hill.

Answers to Selected Problems

1.2 $N_\phi = -p_oR/2$

$N_\theta = -p_oR(\cos^2 \phi - 1/2)$

1.3

$$N_\phi = \frac{-\gamma R^2}{6}\left(\frac{3H}{R} + 1 - \frac{2\cos^2\phi}{1+\cos\phi}\right)$$

$$N_\theta = \frac{-\gamma R^2}{6}\left(\frac{3H}{R} - 1 - \frac{6-4\cos^2\phi}{1+\cos\phi}\right)$$

1.6 $t = 0.90$ inch

1.8

$$\max N_s = \frac{-847\gamma L^2 \sin\alpha}{432} \quad \text{at } s = \frac{L}{12}$$

$$\max N_\theta = \frac{\gamma L^2}{4}\sin\alpha \quad \text{at } s = \frac{L}{2}$$

2.4 $A = 5.76$ inch^2

2.5

$t_1 = 0.21$ inch, $t_2 = 0.82$ inch

$t_3 = 0.98$ inch, $t_4 = 0.69$ inch

$A = 1.22$ inch^2

3.3 max $M_x = 0.322\ Q_o/\beta$ at $x = 0.61\sqrt{rt}$

3.4 At section a–a, $M_o = 0$ and $H_o = 0.0195D\beta^3$

Stress in ASME Pressure Vessels, Boilers, and Nuclear Components, First Edition. Maan H. Jawad.
© 2018, The American Society of Mechanical Engineers (ASME), 2 Park Avenue,
New York, NY, 10016, USA (www.asme.org). Published 2018 by John Wiley & Sons, Inc.

3.5 $M_a = 14.95/\beta$ and $M_b = 44.97/\beta$
4.1 $t = 5/16$ inch
6.2 $t = 0.73$ inch
6.3 $p = 148.9$ psi
7.2 $\sigma_x = 92.2$ MPa, $\sigma_y = 124.9$ MPa
8.3 $= 8670$ psi, max $w = 0.23$ inch
8.4 $= 12,330$ psi
8.6

$$M_r = \frac{p}{16}(3+\mu)(a^2-r^2) - \frac{pb^2}{4}(K_1 + K_2 - K_3 - K_4 - 1)$$

where

$$K_1 = (1+\mu)\left(\frac{3+\mu}{2(1+\mu)} - \frac{b^2}{a^2-b^2}\ln\frac{b}{a} - \frac{1}{2}\right)$$

$$K_2 = \frac{1-\mu}{r}\left(\frac{1+\mu}{1-\mu}\frac{a^2b^2}{a^2-b^2}\ln\frac{b}{a}\right)$$

$$K_3 = (1+\mu)\ln\frac{r}{a}$$

$$K_4 = \frac{(3+\mu)}{4}\left(\frac{a^2-r^2}{r^2}\right)$$

$$M_t = \frac{p}{16}\left[a^2(3+\mu) - r^2(1+3\mu)\right] - \frac{pb^2}{4}(K_1 + K_2 - K_3 + K_5 - \mu)$$

$$K_5 = \frac{(3+\mu)}{4}\left(\frac{a^2+r^2}{r^2}\right)$$

9.1 $M_p = pL^2/8$
9.3 $M_p = pL^2/144$
9.8 $M_p = 157.3\, p$

Index

Stress in ASME Pressure Vessels, Boilers, and Nuclear Components, First Edition. Maan H. Jawad.
© 2018, The American Society of Mechanical Engineers (ASME), 2 Park Avenue,
New York, NY, 10016, USA (www.asme.org). Published 2018 by John Wiley & Sons, Inc.